EMBEDDED INTERNET DESIGN

EMBEDDED INTERNET DESIGN

Al Williams

McGraw-Hill
New York · Chicago · San Francisco · Lisbon
London · Madrid · Mexico City · Milan · New Delhi
San Juan · Seoul · Singapore
Sydney · Toronto

The McGraw·Hill Companies

Cataloging-in-Publication Data is on file with the Library of Congress.

Copyright © 2003 by The McGraw-Hill Companies, Inc. All rights reserved. Printed in the United States of America. Except as permitted under the United States Copyright Act of 1976, no part of this publication may be reproduced or distributed in any form or by any means, or stored in a data base or retrieval system, without the prior written permission of the publisher.

1 2 3 4 5 6 7 8 9 0   DOC/DOC   0 9 8 7 6 5 4 3

ISBN 0-07-137436-1

The sponsoring editor for this book was Scott Grillo, the editing supervisor was Caroline Levine, and the production supervisor was Sherri Souffrance. It was set in Vendome by Joanne Morbit of McGraw-Hill Professional's Hightstown, N.J., composition unit in cooperation with Spring Point Publishing Services.

Printed and bound by RR Donnelley.

McGraw-Hill books are available at special quantity discounts to use as premiums and sales promotions, or for use in corporate training programs. For more information, please write to the Director of Special Sales, McGraw-Hill Professional, Two Penn Plaza, New York, NY 10121-2298. Or contact your local bookstore.

This book is printed on recycled, acid-free paper containing a minimum of 50% recycled, de-inked fiber.

*For my son Patrick,
and always for Pat*

CONTENTS

Contents

ACKNOWLEDGMENTS

Like most big projects, a book is a collaborative effort among people. Although you think of books as having an author, in truth, they are really created by an entire team of people. A book simply wouldn't be possible without each one of them.

To that end, I'd like to express my appreciation to Scott Grillo, Caroline Levine, Sherri Souffrance, Joanne Morbit, and everyone at McGraw Hill for their great work and patience.

In addition, thanks to the folks at Parallax, Maxim, Dallas, and Cermetek for their assistance in putting all of this together. In a very literal sense, this book wouldn't be what it is without the products from these companies.

Last, but certainly not least, my family—especially my wife Pat—is so understanding and supportive of these projects. I couldn't do what I do without her or the rest of our clan and their understanding nature.

INTRODUCTION

If you ever think about the modern telephone system, it is nothing short of a miracle. On my desk I have a box (my phone) with wires coming out of the back of it. By pushing a few buttons, I can connect those wires (virtually, anyway) to any other phone in the world. That's billions of phones. Supporting all of this is miles and miles of copper wire, cable, and radio links.

The Internet has evolved to where it is essentially a phone network for computers. I don't know about you, but if my PC is disconnected from the Internet, I feel like half of it is missing. Although recent years have seen the Internet exploited for users, we are going to see explosive growth in the connection of embedded devices to the Internet.

There are several reasons for this. First, embedded devices are already everywhere. In the 1980s we talked about the day when every home would have a computer. These days, although many homes do have a personal computer, practically every house has large numbers of computers embedded into things.

Toys, telephones, televisions, VCRs, DVD players, answering machines, stereos—almost anything that plugs into the wall has an embedded system in it these days. Even remote controls usually have a microprocessor. Most cars now have several microprocessors along with a growing number of portable devices like PDAs and cell phones.

Just as PCs are more useful and powerful when they are connected together, embedded systems also benefit from connectivity. With the pervasiveness of the Internet, it makes sense that embedded systems will connect—with or without wires—to the Internet to communicate with other embedded systems and even PCs (and, through PCs, to users).

In the home, you might use a cell phone to turn the heat on before arriving home from a trip. In industry, a temperature sensor might e-mail daily status reports, provide a Web page with current data, and alert instant messenger users when a temperature is out of range. It is easy to see how powerful a connected embedded system can be.

Today, connecting a simple microcontroller to the Internet can be a bit tricky. Most small microcontrollers don't have a way to connect to a network. Beyond that, there are many ways any computer might connect to the Internet: phone modems, Ethernet, USB DSL or cable modems, or a serial port connection to a larger computer.

This book looks at several ways to build embedded Internet systems. You could use most of the devices you'll see in this book to build a complete system or you could use them as an Internet interface for another microcontroller.

Microprocessors revolutionized the computer business forever. The Internet transformed personal computing and communications in ways most of us never imagined. The convergence of the Internet and embedded systems promises to change our world yet again, and probably in unexpected ways.

The exciting part is that this technology is nascent—we are just starting to explore what it means to have inexpensive, ubiquitous connectivity between embedded systems, PCs, and humans. There's still plenty of room for innovation and discovery. This book is your gateway to making this important connection.

Is This Book for You?

Creating an Internet-enabled system requires a blend of skills beyond the traditional hardware skills. Internet programming can be a substantial task since you have to understand the low-level network protocols (like PPP, for example) and you also have to understand the high-level network programming (for example, how can you send e-mail or request a Web page).

One language, Java, has been a leader in making the high-level networking simple, and most of the programs in this book will use some form of Java or a Java-like language to take advantage of this fact. The TINI—a tiny embedded processor—is a Java powerhouse that has capabilities rivaling a desktop computer. The Parallax Javelin uses a language that is based on Java. Although it isn't as powerful as a desktop computer, it does things that traditional desktop computers have difficulty doing (like making precise time measurements, generating and reading analog voltages, and similar real-time control tasks).

To get the most from this book, you don't need specific experience with these processors. However, you should be familiar with the ideas behind embedded programming. Experience with a high-level language (although not necessarily Java) will be useful, too. This book will show you in detail how to utilize several different processors to send and receive e-mail, serve Web pages, and connect to the Internet using a variety of mechanisms. You'll find projects that can form the basis of your own designs as well as practical advice about which tools will best fit your application.

What's Inside?

This book can be segregated into three major sections. In the first section, you'll find valuable background information, a quick introduction to Java, and some Java software that can turn an ordinary PC (or other desktop computer) into a bridge for any RS-232 capable microcontroller. The second section focuses on the Maxim/Dallas Semiconductor TINI, which is a small Java-based microcontroller with impressive networking capabilities. Finally, the third section shows you how to use the Parallax Javelin (a small controller that runs a variation of Java) with different types of modems. Here's the breakdown by chapter:

Chapter 1. This chapter explains how the Internet works under the hood and what's required to successfully communicate with Internet hosts.

Chapter 2. Although Java has a reputation for being difficult to learn, most of that is due to graphical operating system concerns. In this chapter, you'll see that for embedded systems, Java can be quite simple to learn and use.

Chapter 3. To practice with Java, you can examine this PC-based Java program that bridges between the Internet and a Basic Stamp.

Chapter 4. If you want to embed a powerful Java-based processor, the TINI is a great solution, as you'll see in this chapter.

Chapter 5. The TINI is powerful enough to act as a Web server. This chapter shows you how to connect to the TINI using any Web browser.

Chapter 6. This chapter introduces the Parallax Javelin, which is an easy-to-use microcontroller that runs a language very much like Java.

Chapter 7. Parallax provides classes for the Javelin to connect to the Internet via a modem. This chapter shows how to use these classes.

Chapter 8. This chapter dives inside the Parallax network stack so you can understand what's going on and modify things if necessary.

Chapter 9. You'll find a Javelin using a Cermetek iModem to send e-mail in this chapter. The iModem is typical of a class of ready-to-use Internet appliances aimed at the embedded market.

Chapter 10. Wrapping things up, I'll show you some devices we didn't talk about in the rest of the book as well as future developments that could impact embedded developers.

Using This Book

Unlike some books, you probably won't read this one in chapter-by-chapter order. Probably everyone will want to read Chapter 1. If you are already a Java guru, you can safely skip Chapters 2 and 3. Otherwise, it is important to read these chapters so you can follow the code in the subsequent chapters.

After those first three chapters, you'll probably want to jump to the processor that interests you the most. If you are looking for an Ethernet connection and the kind of Java language you traditionally find in a PC, you'll want to dig into the chapters about the TINI (that is, Chaps. 4 and 5). If you want lower-level control and PPP connections over a modem, read up on the Parallax Javelin in Chaps. 6, 7, and 8. If you'd like to connect an off-the-shelf iModem to any processor, you'll want to pay special attention to Chap. 9.

If you just want an overview of how this important technology will work, you might read each chapter in order (even Chap. 8, which is somewhat detailed, offers valuable insights into PPP).

Whatever course you elect to take, it is sure to start in the same place. So flip over to Chap. 1 and you'll be that much closer to adding the Internet to your embedded designs.

AL WILLIAMS

1

Under the Hood

Predicting the future is a dangerous game. In the '80s, computer professionals (and hobbyists) spent a lot of time debating the merits of "the home computer." The consensus was that—one day—everyone would have a computer in their home. You'd use it to balance your checkbook, prepare recipes, and play games.

Truth is stranger than predictions. Today, many people do have a personal computer in their homes, but by far the biggest single use of it is to connect to the Internet. Sure, some people do use PCs for other things, but most bought a computer purely to access the Internet.

On the other hand, most homes have plenty of computers in them—not necessarily PCs, but computers nonetheless. Today, you'll find dedicated computers inside televisions, telephones, microwave ovens, alarm clocks—you name it. Some microcontrollers are easy to spot—video games, laser printers, even computer mice. Others might surprise you—thermostats, washing machines, and garage door openers may even contain processors.

How did this happen? Why does something as simple as an alarm clock need a computer? The answer lies in economies of scale. Sure, you could design an alarm clock that used dedicated logic—you'd need an oscillator for the time base and a series of counters. You'd also need to decode the count and drive light emitting diode (LED) or liquid crystal display (LCD) digits. Of course, you'd also have to handle the alarm functions, so you'd need another counter, a comparator, and a slew of switching logic.

A clock like this might work, but it wouldn't be economical for you to produce. Besides, it would probably be the size of a loaf of bread! The next step—and the one that made digital clocks practical—is for a semiconductor company to do the same design, but on an IC instead of using discrete logic gates. The semiconductor company sells the clock chip to many clock companies, and therefore sells more chips. This allows the semiconductor producers to spend more time on the design than any one clock company could probably afford.

What if the semiconductor company wanted to sell even more chips (either to make more money, to drive down the price of the chips, or both)? It might decide to make the chip support clock functions by providing general-purpose building blocks. Instead of making a clock on a chip, you could make a chip that has a time base, a counter, a display driver, and so forth. If you connect a few pins together, you've built a clock. If you add an input buffer, and wire some different pins, you've built a frequency counter. Add another connection, and you have a counter or a speedometer.

Now clock makers will buy the chip, but so will instrument manufacturers. Perhaps the chip will find its way into some cars, or into a stopwatch design. As more people use the chip, the company sells more chips and—because of economies of scale—it can reduce the price.

The microcontroller is the ultimate extension of this idea. Instead of making a specific chip (like a clock), it makes more sense to design a microcontroller. You can think of a microcontroller as being a chip full of building blocks. The designer connects these building blocks with software, not with wire. One type of microprocessor might wind up in a pocket calculator, a TV remote control, an airplane's guidance system, and a specialized piece of lab equipment. Each piece of equipment does a different job, but a programmable chip can do each job. Since so many of these chips will be sold, even though they are more complex devices than dedicated chips, they are less expensive to make, and thus to sell. Some microprocessors sell for well under a dollar in quantity.

A Marriage?

Microprocessors are ubiquitous and the Internet is pervasive. Then it is no surprise that many developers are connecting their microprocessors to the Internet. In theory, we are about to enter a brave new world where you can program your VCR at home from your Web browser at home. Left the coffee pot on? Dial up the Internet on your Web-enabled cell phone and turn it off. Let your pool e-mail you when the pump needs backwashing.

Of course, the devil is in the details. Today, your coffee pot probably doesn't have a connection to the Internet. Some consumer devices (like satellite receivers, for example) require a phone connection, but that won't work for devices needing constant access. Clearly, tomorrow's home will have some sort of network built-in.

Developer Profile

The other problem is in the skill set required to develop devices like this. Traditionally, microcontroller designers concentrated on hardware and low-level software. Internet developers usually work with high-level constructs and rarely worry about the low-level workings of the Internet.

To successfully design a microcontroller-based Internet device, you may find you have to get down and dirty with hardware and low-level problems. But you'll also need to handle high-level ideas like Hypertext Transfer Protocol (HTTP) and e-mail formats.

For example, suppose you want to design an Internet-based thermostat to control a home heating system. You could design custom software to talk to the thermostat. However, that doesn't make much sense. Why reinvent the wheel when companies like Microsoft and Netscape spend small fortunes perfecting their Internet software? It would be smarter to make the thermostat a micro Web server. You could monitor and control the thermostat via any Web browser anywhere in the world.

To successfully design an embedded Internet device, you need to understand all of the following:

Hardware design

Programming

Low-level network protocols

High-level Internet protocols

If you are reading this book, it is a good bet you regularly surf the Web and get e-mail. However, do you know about how those things really work at the lowest levels?

Inside a Web Session

Suppose you sit at your personal computer (PC) and call up a Web page—perhaps Yahoo, for example. When you type in the universal resource locator (URL) http://www.yahoo.com into your browser, you are creating an HTTP request. The request superficially resembles a file name, with several important differences. In addition to the URL, the browser may send other data in the request (for example, form data you've entered, cookies that the server set earlier, the browser's type, and your preferred language).

The server processes this HTTP request and forms a response. The response, in this case, is the Yahoo home page. Both the request and response have two distinct parts. The first part is the header. This section contains information about the request or the response. For example, a header might indicate the language used, the server or browser version number, or the type of data the browser is willing to accept.

The second part of the data is the actual content. For a request, this is often blank. However, if you are submitting a form to the server, the data may appear in the content (although there are other places it might appear as well). For the response, the content is the Web page or other served content. For example, for a picture the content would be the data that comprises the Graphics Interchange Format (GIF) or Joint Photographic Expert Group (JPEG) file.

Making a Connection

The browser (known as the *client*) sends an HTTP request to the server and the server replies with a response. This leaves the question: How do the client and the server actually communicate with each other?

On the Internet, the answer is Transport Control Protocol/Internet Protocol (TCP/IP). This is a network protocol that works using sockets. Both clients and servers use sockets. Each machine has an IP address—this is a 4-byte number that uniquely identifies the machine on the network—like a phone number. You often see these addresses written as four decimal numbers separated by periods, as in 10.24.128.1. Since each decimal number represents a byte, the value of each portion (an octet) will range between 0 and 255.

Of course, people don't want to remember some cryptic IP address—they want a human-readable name. For that reason, the Internet is full of domain name service (DNS) machines. These servers allow computers to look up a name (like www.yahoo.com) and resolve it into an IP address. Your browser does this automatically.

Each machine on the network can have a many ports. These ports are how the computer talks to other computers. Each port has a number. Ports numbered between 0 and 1023 are used for well-known servers. For example, Web servers use port 80, while e-mail servers use port 25. In addition, port numbers 1024 to 49151 are registered to certain programs by the Internet Assigned Number Authority (IANA). These are generally special-purpose servers, and are not as widespread as the well-known servers. Ports numbered above 49151 are available for any use.

When a client computer wants to connect to a server, it needs its IP address and the port number for that particular service. Your browser automatically resolves the host name you enter (using DNS) and supplies port 80 (by default). Using this information, the client creates a socket (its port number doesn't matter) that will connect to the server's port.

The server computer, of course, must be running Web server software. This software creates a socket on port 80 to listen for connections. However, if a Web browser actually connected to port 80, it would tie up the server and make it unavailable for other clients. To avoid this situation, the server port behaves in a special way. The server does listen on port 80—however, when a client attempts to connect on this port, the server creates a new port (with some random port number) to handle the request. This leaves the original port 80 open for further connections. While the client thinks it is using port 80, it is really using a different port, but the result is the same.

You can actually impersonate a Web browser using a telnet program. (You usually use a telnet program to remotely log in to a computer over the Internet.) Depending on your telnet program, you can usually specify a different port to use instead of the default telnet port (port 23). If you are using Windows 98, or a similar operating system, the telnet program allows you to override the port on the connection dialog. For most operating systems (including Windows 98 and Windows 2000), you can simply specify the port on the command line, as in the following:

```
telnet www.yahoo.com 80
```

Depending on your telnet program, you may not be able to see what you type, but you can enter something like

```
GET /
```

When you press the Enter key, the Web server will respond with the Web page you asked for (in this case, the default Web page) and disconnect you. If you can scroll back (or if you are a very fast reader) you'll see the headers, a blank line, and the content of the Web page. Figure 1-1 shows a sample telnet session connected to a Web server.

All Web applications use sockets in this manner. The biggest difference between a Web browser, a telnet client, and an e-mail program is what the client and server say to each other. The underlying communications mechanism is the same.

More about IP Addresses

For a client, any IP address will do. After all, the client connects to the server, so the server doesn't need to know the client's address any more

Figure 1-1
Using Telnet to Connect to a Web Server

```
GET /
HTTP/1.0 200 OK
Content-Length: 16170
Content-Type: text/html

<html><head><title>Yahoo!</title><base href=http://www.yahoo.com/><meta http-equ
iv="PICS-Label" content='(PICS-1.1 "http://www.rsac.org/ratingsv01.html" l gen t
rue for "http://www.yahoo.com" r (n 0 s 0 v 0 l 0))'></head><body><center><form
action=http://search.yahoo.com/bin/search><map name=m><area coords="0,0,52,52" h
ref=r/a1><area coords="53,0,121,52" href=r/p1><area coords="122,0,191,52" href=r
/m1><area coords="441,0,510,52" href=r/wn><area coords="511,0,579,52" href=r/i1>
<area coords="580,0,637,52" href=r/hw></map><img width=638 height=53 border=0 us
emap="#m" src=http://us.a1.yimg.com/us.yimg.com/i/ww/m5v2.gif alt=Yahoo!><br><tab
le border=0 cellspacing=0 cellpadding=3 width=640><tr><td align=center width=205
>
 <a href="/homet/?http://mail.yahoo.com"><b>Yahoo! Mail</b></a><br>free email fc
r life</td><td align=center><a href="http://rd.yahoo.com/M=89335.886962.2641560.
389576/S=2716149:NP/A=421897/*http://sweepstakes.bluelight.com/promos/backtoscho
ol/?c=MKTGYH338080400" target="_top"><img width=230 height=33 src="http://us.a1.
yimg.com/us.yimg.com/a/pr/promo/bluelight6/hp_bluelight1.gif" alt="Win with Blue
light.com!" border=0></a></td><td align=center width=205><img src="http://us.i1.
yimg.com/us.yimg.com/i/new2.gif" height=11 width=28> Pro Football<br><a href="/h
omet/?http://football.fantasysports.yahoo.com/full"><b>Fantasy</b></a> - <a href
="/homet/?http://football.fantasysports.yahoo.com/pickem"><b>Pick'em</b></a></td
></tr><tr><td colspan=3 align=center><input size=30 name=p>
```

than you need to know the phone number of someone who calls you. However, a server always needs a well-known IP address (or at least a well-known name that DSN can resolve).

How does a machine get associated with an IP address? That's a complicated question. The short answer is that one of several regional authorities (using guidelines set by the Internet Assigned Number Authority) assigns IP addresses to organizations. There are three types or classes of IP address—class A, B, and C. You can tell the class of an address by the numbers at the beginning of the address (see Table 1-1). However, due to the rapid depletion of IP addresses (because of the explosive growth of the Internet), some schemes are in use to further subdivide IP addresses. So one organization may own a part of a class B address while another company uses another part of the same class B network address.

In real life, most people never deal with addresses at this level. Usually, your ISP, your company, or your company's Internet service provider (ISP) will have an IP address (or a block of IP addresses) assigned to it. The organization might preassign addresses to specific machines. This is known as a static IP address and it is almost a necessity for server machines. Many ISPs charge extra to assign static IP addresses.

A more common scheme is to use Dynamic Host Control Protocol (DHCP). This allows a client machine to acquire an IP address dynamically. If you use a dialup connection, you almost certainly have a dynamic IP address. In other words, each time you connect to the Internet, your ISP assigns you an address from a pool of addresses. You might get the same address you had last time, but the odds are against it.

Class	Prefix range	Maximum addresses in class	Theoretical maximum addresses per network	Example	Also known as
A	1–126	126	16777214	51.120.7.99	/8 address
B	128.0–191.255	16384	65535	180.16.40.1	/16 address
C	192.0.0–223.255.255	209712	254	192.10.1.1	/24 address

TABLE 1-1 IP Address Classes

This poses a special problem for embedded systems. Suppose you design the Internet-based thermostat I mentioned earlier. If the thermostat can have a dedicated IP address, then you could easily make it a Web server. However, if the thermostat's IP address changes frequently, you'll have to find some way around the limitation.

Floating IP

Dynamic IP addresses are one reason it is difficult to run any sort of server on a dialup connection. Of course, most people using dialup connections are only surfing the Web—no one connects to them, right? That might be true if it were not for e-mail. It would be very inconvenient if people could only send you e-mail while you were connected to the Internet. Of course, this isn't a problem. Why? Your ISP maintains a Post Office Protocol 3 (POP3) [or Internet Message Access Protocol (IMAP)] server that fields incoming e-mail for you. When you log in, your e-mail program checks the POP3 or IMAP server for any mail waiting for you. Sending mail, of course, is no problem, since you'll always connect to the Internet to send mail via an SMTP server—again provided by your ISP.

This suggests one possibility for an embedded system: e-mail. As odd as it may sound, you may want to build your Internet thermostat so that it sends and receives e-mail! That solves many problems since the thermostat can poll for incoming messages when it is ready to process them. It also solves any problem with varying IP addresses. Even if the thermostat connects intermittently via a modem, it can still check for queued e-mail.

Of course, the downside is that a server—perhaps a server provided by your ISP—must exist to handle the incoming e-mail requests. However, since many ISPs provide this server anyway, you might as well use it.

You can't always use e-mail, so you may have to devise other schemes to work around dynamic IP addresses. For example, you might have the device store its address on another server that has a permanent IP address. Other programs would have to retrieve the IP address from this main server before contacting the device.

Decisions

Assuming you want to connect to a remote device, you'll have to choose what method you use. The most flexible option is to simply use raw sockets to connect the server (usually the embedded device) to a client program. The client program might be running under the control of a user, or it might be running on another server (for example, a Web server) to provide data for other purposes.

Of course, this requires you to write custom client software. Often, you won't want to develop a complex client. If your server looks like a Web server, you can hijack a client that is on nearly every computer—a Web browser. It would be very difficult to write software for your own client with even some of the capabilities of a browser. Also, browsers are available for nearly every type of computer you can imagine. Making your embedded system a Web server allows you to use practically any type of computer—a Mac, a PC, a Unix System, or maybe even a Web TV—to monitor and control the device.

Another alternative is to use e-mail. This allows existing e-mail programs to send and receive messages, although communications may not be immediate. This approach also allows you to avoid writing client software. There are many e-mail programs that exist for many computer systems and you won't have to reinvent the wheel.

There are many other Internet protocols you might consider using for special purposes. For example, you might have a device that sends data files via File Transfer Protocol (FTP) to an FTP server. However, for most projects you'll want either a custom socket or a Web-based solution.

URL Details

The core way you specify a Web address is via a URL. You've certainly used URLs before, but there are a few special parts of the URL you may find useful that you don't use often as an end-user.

Internet Documentation

Unlike a corporate network, the Internet is somewhat chaotic, with a very loose control structure. This dates back to the late 1960s when the U.S. government established the Internet as a network connecting several important research centers and universities. Most major Internet protocols—including HTTP, e-mail formats, and the like—are codified in requests for comments (RFCs).

When a person (or group) wants to make a new standard known, they may write and submit an Internet Draft (ID) document. These drafts are collected and housed by the Internet Engineering Task Force (IETF). Once a document is in draft form, you can request that the RFC editor consider the document for publication as an RFC.

Not all RFCs are standards—some define preliminary or experimental protocols and formats. However, there are many very important RFCs that define the Internet you use everyday. For example, RFC1123 defines services provided by Internet hosts (and refers to many other RFCs). Not all RFCs define something. RFC1207, for example, is a frequently asked question list for advanced Internet users.

If you create a widely applicable protocol for communicating with devices, you might consider publishing your own RFC. Some tongue-in-cheek RFCs from the past in this vein include RFC2324 (Hyper Text Coffee Pot Control Protocol HTCPCP/1.0) and RFC2325 (Definitions of Managed Objects for Drip-Type Heated Beverage Hardware Devices using SMIv2). I have little doubt that RFCs that really define protocols for controlling devices will appear soon.

One thing to watch out for when reading RFCs is that they can be out of date. The reason is that an RFC—once published—can never change. Any serious errors or improvements require a new RFC. For example, RFC822 defines e-mail formatting. However, when the need to send attachments and nontext e-mail arose, it took a new RFC (RFC1341) to handle the new content types.

Consider this rather complex URL:

```
http://joe:zippy@www.al-
williams.com:8000/somedir/lang.htm?v1=hello&v2=joe#done
```

Here is what each part of this URL indicates:

http:// The protocol to use. HTTP is the normal Web protocol. However, you can also use ftp://, file://, or other legal protocols.

joe:zippy. This is the user (joe) and password (zippy). This is only required, of course, for pages that are password protected and require user authentication.

www.al-williams.com. This portion of the URL identifies the server that contains the page. By default, the browser will connect to the server using port 80.

8000. This indicates that the browser should connect using port 8000 instead of the default port, 80. Sometimes, a machine will run more than one server, and use different ports for each server. For example, one server might be for public Web access, but another server—on the same machine—monitors for administrative requests, or for signals from remote sensors.

somedir. The document's directory on the server machine. This may consist of more than one subdirectory, separated by slashes. The directory is relative to some root directory on the server machine as determined by the Web server. For example, on a Windows host, the files may be relative to c:\inetpub\wwwroot.

lang.htm. The file name directs the server as to which document to load. Usually, this is simply a Hypertext Markup Language (HTML) or text file. However, in some cases it may be a program that the server executes to generate the output. If you don't specify a file name, the server will either select a default document (usually index.htm or index.html) or—if the server is configured to do so—it will display a directory listing of the specified directory.

v1=hello&v2=joe. The portion of the URL after the question mark is the query string. In this case, there are two variables *v1* and *v2*. Remember, not all characters are legal to use in a URL, so certain characters require special handling. Spaces, for example, should appear as plus signs. You can encode any character by using a percent sign, followed by a two-digit hex number. This number is the character's American Standards Committee for Information Exchange (ASCII) code. For example, you can encode a space as %20. You can represent any character in this way. For example, an uppercase A—which doesn't require special encoding—may be written as %41.

done. Text named after the # symbol causes the browser to jump to a named label (called an *anchor*) within the document.

You'll rarely use all of the parts in a single URL, but you can. The simplest URL only contains the protocol and the server name. As an end-user, you may never find it important to use the user name, password, port, or query string directly. However, when working with embedded systems, it is often useful to direct your requests to an alternative Web server, or to pass data in the query string.

All about Content

Sockets are the fundamental way that programs communicate over the Internet. Even if you want to use e-mail, or a Web browser, you still have to construct and connect a socket. That takes care of the underlying connection between your embedded device and the remote computer. But once you establish that connection, what do you say?

Earlier, you saw that Web transactions (and e-mail) are divided into headers and content. The headers contain information about the document (the date it last changed, or its size, for example). This begs the question: What is in the content section?

One header defines the content's type (appropriately, the *content-type* header). That means the data in the content section might be anything. For example, it might be a GIF or JPEG file. However, the most common content type is HTML.

HTML is the *lingua franca* of the Web. If you've ever written a Web page, you've used HTML. You'll find more information about HTML in Appendix A, but the following will give you enough background to get through most of the HTML you might need.

HTML files look like ordinary text files, but they contain special commands known as *tags*. A tag is some text enclosed in angle brackets, like *<BODY>*. Some tags have corresponding ending tags (for example, *</BODY>*) but others stand alone. Some tags have one or more optional attributes. For the tag *<BODY BGCOLOR=BLUE>*, the attribute is *BGCOLOR* and the value of the attribute is *BLUE*. If the value contains special characters like spaces, you should enclose it with single or double quotes. Of course, you can always use quotes, so the previous example could have been *<BODY BGCOLOR="BLUE">*. However, a tag like *<A NAME="Document Top">* must use quotes in the attribute because of the space in the value. Tags are not case-sensitive. The tags *<BODY>*, *<body>*, and *<bODy>* are all the same.

You can encode characters using the ampersand character (&). For example < is a less than sign. This is an important character because HTML treats the less than sign as the start of a tag. If you really want the < character to appear, you must use < instead. Another special character is the ampersand itself, which is *&*. You can encode any character by using *&#nnn;* where *nnn* is the decimal number that corresponds to the character's ASCII code. Some of the character names are case-sensitive (unlike tags).

Anatomy of a Document

Each HTML document has two main parts: a head, and a body. The head is similar to, but not the same as, the document headers. The head contains information about the document (although, for the purposes of the socket communications, the head is part of the content). For example, your browser probably shows a page title in the title bar. This title appears in the head of the document.

The body is the actual text, along with information like graphics, hyperlinks, and forms. The browser will format the text, as you'll see shortly. It also acts on special formatting commands embedded in the text. These commands also delineate the sections of the document.

One frustrating thing about HTML is that it is up to the browser to interpret the document. That means you don't have complete control over how the page will look. It also means that one browser may show a page much differently than another browser. It also means that a document that one browser will accept may produce an error in a different program.

Although many HTML documents don't include it, the first line is supposed to contain a special comment that identifies the version of HTML the file obeys. However, you don't always find this special line, and browsers don't really care if it is missing.

The next thing in the file (or the first thing, if the special comment is missing) is a special tag that marks the file as an HTML file. Tags are how you specify commands inside your HTML document. Remember, tags always appear between the < and > characters. In this case, the tag is *<HTML>*. Many tags, including the HTML tag, appear in pairs. That means that at the end of the file, there must be a corresponding *</HTML>* tag. Listing 1-1 shows a very simple HTML file that includes the special version number comment, an HTML tag, and a BODY tag. The BODY tag, of course, marks the start of the document's body section.

This is a very simple file:

```
</BODY>
</HTML>
```

```
<!DOCTYPE HTML PUBLIC "-//W3C//DTD HTML 4.0 Transitional//EN"
  "http://www.w3.org/TR/REC-html40/loose.dtd">
<HTML>
<BODY>
```

In the document in Listing 1-1, there is no head section. That isn't illegal, but it does mean the browser will set the document's title and other information arbitrarily. Notice that the <!DOCTYPE> tag specifies a particular flavor of HTML defined at the specified URL. In this case, the definition is the relaxed (transitional) HTML 4.0 specification for the English language.

Listing 1-2 shows a more complete Web page. Notice that the head section is present (before the body) and contains a title tag. In addition, the body now contains a P tag. This indicates a paragraph, and is crucial for controlling line breaks.

If you want to add comments to an HTML file, you can use a special syntax. Start your comment with <!-- and end it with -->. These comments will not appear in the browser's window, but will be in the HTML file to explain or document anything you feel is important to people viewing the file directly.

Formatting

Without any special commands, the browser reformats any text to fit the browser's window. To do this, it collapses any consecutive white space characters into a single blank. It then wraps lines as necessary by replacing selected blanks with new line characters.

That's why the P (paragraph) tag is so important. Without it, your entire document would just be one long paragraph. Technically, the P tag requires a closing tag, but sometimes authors omit the closing tag since browsers will still correctly interpret documents that only have the opening paragraph tags.

Another way to force a line break is with the BR tag. There is no closing tag for BR. It doesn't start a new paragraph; it simply breaks the current line, which isn't quite the same thing. For example, any spacing between paragraphs doesn't apply to a BR tag.

```
<!DOCTYPE HTML PUBLIC "-//W3C//DTD HTML 4.0 Transitional//EN"
  "http://www.w3.org/TR/REC-html40/loose.dtd">
<HTML>
<HEAD>
<TITLE>A more complete file</TITLE>
</HEAD>
<BODY BGCOLOR=White>
<P>This file is a bit more complete. It has a head section with a title.
<P>It also has more than one paragraph.
</BODY>
</HTML>
```

Listing 1-2
A More Complex HTML File

Sometimes you will want the browser to obey your particular spacing and line breaks exactly. In that case, you can use the *PRE* tag (and, of course, the corresponding ending tag). This will cause most browsers to display your text in a fixed-pitch font and obey your spacing even if it results in a page too wide for the user's browser to display without scrolling.

Formatting the appearance of text is straightforward, but the browser is not required to obey. A browser might run on a computer where it is not possible to show italic text, for example. For this reason, there are two ways to format text. You can specify a specific format (like bold or italic) and hope that the browser can display it. Alternatively, you can specify more generic formatting codes—for example, emphasis (*<EM>*), or strong emphasis (*<STRONG>*). The browser decides how to display those elements, and you can depend on it being consistent within a single browser. Table 1-2 shows a list of formatting tags that you can use.

Forms

You usually think of Web servers as sending data to the browser. However, there are times when a browser (or any client) might want to send data back to the server. For example, you might enter a search phrase to

TABLE 1-2

Common Formatting Codes

Tag	Definition
<B>	Boldface
<I>	Italics
<U>	Underline
<TT>	TeleType (monospaced)
<CITE>	Book citation
<CODE>	Source code
<DFN>	Definition of a word
<EM>	Emphasis
<KBD>	Text a user should type
<SAMP>	Sample output
<STRONG>	Strong emphasis
<VAR>	Variable placeholder

a search engine. A remote sensor might want to inform the server about its current reading and status.

When a Web page (or any Web client) wants to send data back to the server, it has two choices. First, the data may appear in the URL's query string (the portion after a question mark). For example, http://www.al-williams.com/cgi-bin/data.pl?temp=200 would send the data *temp*=200 to a script on my Web server. This is often known as a *GET* since the format of the data is the same as the way a Web user requests a page. In fact, a user could type in the correct URL to send data to the server without a form.

The other way to send data to the server is via a POST transaction. With a POST, the data doesn't appear in the URL, but is sent in the content portion of the request. In either case, a normal Web page uses a form to collect the information from a user. When the user submits the form (usually using a Submit button), the browser sends the data to the server.

Look at Fig. 1-2, which shows a simple form. Listing 1-3 shows the HTML that creates this form. Each line has a particular function:

<FORM...>. This line introduces the form. The *ACTION* attribute indicates the URL for the server that will accept the form's data. The *METHOD* attribute indicates how to transmit the data.

Figure 1-2
A Simple HTML Form

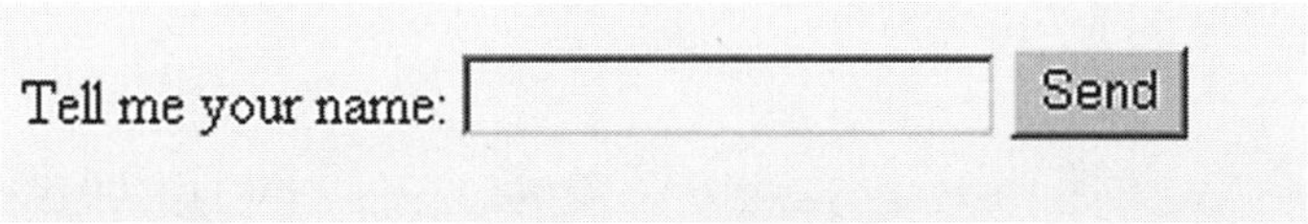

Tell me your name: [   ] Send

Listing 1-3
A Sample Form

```
<HTML>
<HEAD>
<TITLE>A Simple Form</TITLE>
</HEAD>
<BODY>
<FORM ACTION=test.asp METHOD=post>
Tell me your name: <INPUT TYPE=TEXT>
<INPUT TYPE=SUBMIT VALUE=Send>
</FORM>
</BODY>
</HTML>
```

<INPUT TYPE=TEXT...>. Each *INPUT* tag defines a data field. The *NAME* attribute defines the field's identifier. The value of the field will be what the user enters.

<INPUT TYPE=SUBMIT>. This tag defines a button that will cause the browser to send the form's data to the server.

</FORM>. The *FORM* tag simply indicates the end of the form.

Of course, there are many other elements you can place on a form. For our purposes, these simple elements will suffice. The server, of course, must have some sort of program to accept and interpret the data from the form. Traditionally, Web servers process data using a common gateway interface (CGI) program. This is just a program that accepts data from a form and can produce HTML output. These programs are often written in Perl—a popular interpreted language—but there is no reason they can't be written in most other languages including C or C++.

Modern servers often have special ways to embed programming commands into Web pages. For example, Microsoft's servers allow you to write active server pages (with an .ASP extension) that can work with form input. Other servers may allow you to use PHP or Java server pages (JSP) to do the same sort of manipulation. You can also write special Java programs known as *servlets* to process form data.

XML

Lately there has been a lot of hype about Extensible Markup Language (XML). You might wonder if XML will replace HTML. The answer to that is simply: No. However, XML will have an important role in the Internet and very likely will be useful to embedded developers.

HTML is a language used to describe roughly how a page should appear in the user's browser. HTML makes no distinction between a doctoral thesis and a laundry list. XML, on the other hand, is all about describing data.

XML uses tags similar in structure to HTML's tags. However, in XML you define your own tags, and the syntax rules for XML are more rigorous. For example, in an HTML tag, you only need to use quotes for an attribute if it contains spaces or other special characters. For example, these two hyperlinks are identical:

```
<A HREF=zip.htm>
<A HREF="zip.htm">
```

In XML, the quotes are not optional. Also, XML requires each tag to either have a corresponding ending tag or to end with a /> to indicate that there is no closing tag.

Here is a simple XML file that represents the data from a temperature sensor:

```
<TEMP>
<VALUE>33.2</VALUE>
<UNITS>F</UNITS>
<CALFACTORS VERSION="1.22"/>
<RAW>2010</RAW>
</TEMP>
```

In this example, all the XML tags are custom-made. Because XML tags are made up, programs need guidelines for what constitutes legal data. To define the legal tags (and the sequence of tags), you'll define either a document type definition (DTD) or a schema. A schema is just an XML document that uses special tags to define your custom tags.

If you are sending XML data, you might specify your own schema or you may have to use a schema defined by the recipient. In either case, the recipient will use a special program known as an *XML parser* to pick apart the data. For display purposes, there are several ways to transform XML into HTML. Different devices might format the data differently. For example, a PC-based Web browser might display the temperature on a Web page with color-coding. A Web-enabled cell phone, however, might show only the temperature in a bare-bones format.

Of course, not all recipients will even display the data. An XML client might archive the data to a database, for example. Since XML is infinitely flexible, you can also use it to formulate requests. For example, the same temperature sensor might handle requests in an XML format, as in the following XML request, which uses a single tag to ask the sensor to deliver the current reading (in Fahrenheit) every 10 minutes:

```
<TEMPREQUEST UNITS="F" FREQ="10" IP="192.168.0.1"/>
```

There are a variety of ways that programs can pick apart an XML request. Usually, the recipient will use an XML parser to break the XML into pieces. This parser is a component that works with some other programming language like Java.

The Java Connection

Speaking of Java, Java is a language that has a special relationship to the Internet. Superficially, Java is similar—but not the same as—C++. However, Java's library fully supports the Internet. Creating a socket, for example, requires just a line or two of code. Not only is it simple to work with URLs and sockets, but the way Java programs execute makes it easy for programs to ensure that a program delivered via the Internet will not cause deliberate harm to the user's system.

The two major Web browsers—Internet Explorer and Netscape Navigator—can incorporate small Java programs known as *applets* into a Web page. Because a Java program could wreak havoc on a user's computer, the browsers enforce strict security when running these applets. For example, these applets may not read or write files on the user's machine. They can make network connections, but only to the machine that they originate from.

It is possible for the user to relax these restrictions on applets that contain digital signatures proving that they are from a trusted source. However, you rarely see this done in practice since each browser has a different way of marking an applet as safe.

You'll learn a lot more about Java in future chapters. When writing Internet-enabled programs, Java is the quickest way to get results. Of course, Java requires a certain amount of overhead (both in extra hardware and in execution speed), so you can't always justify its use—especially in high-volume, cost-sensitive production situations. However, if you want to save engineering time, you'll find Java to be the most productive networking language around.

SUMMARY

In this chapter, you've seen that you have to be a bit of a "Renaissance man" to successfully develop hardware that connects to the Internet. Of course, you need to know how to design hardware and write software. But you also need to understand both the high-level and low-level protocols that Internet programs use to communicate. Add to this some knowledge of HTML and XML and you have a good start at developing an entire system.

This book will give you a great start at developing all of these skills. Of course, each of these topics alone could (and, in fact, do) fill an entire book each. You may find you'll need to pick up a few books or read a few Web sites to dive deeper into areas where you need more depth.

ONLINE RESOURCES

www.w3.org. The HTML Specification.

www.htmlhelp.com. A good site for HTML syntax.

www.xml.org. Find out more about XML.

www.faqs.org. Archive of Internet RFCs and FAQs.

www.rfc-editor.org. The official RFC repository.

www.ietf.org. Internet Engineering Task Force

A Java Crash Course

Have you ever dealt with someone who doesn't speak your language? It can be very frustrating. Especially if the other person assumes that if they scream loud enough, you'll understand them.

Regardless of your native tongue, people all around the world use language to communicate in the same way. That's why it is possible to translate from one language to another. Many humans are adept at translating on the fly. Even a few computer programs can translate languages with fair accuracy.

Although human languages are essentially equivalent, that doesn't mean they are equally adept in all situations. Languages that use a small regular alphabet (like English) simplify interaction with computers. Pictographic languages (like forms of Chinese and Japanese, for example) don't lend themselves to simple keyboard entry. On the other hand, pictograms are much simpler if you are writing an optical character recognition program.

Over the years, I've programmed in many different languages ranging from assembly language to many high-level languages. For the most part, these languages are based on the same basic principles. Sure, there are a few exceptions (like Lisp and Prolog). However, languages like Basic, C, Fortran, and PL/I all work in the same basic fashion. Of course, the details of each language can be quite different.

Often, we get comfortable with some language that we like to use and we tend to use it without even thinking about it. I'm like that with C and C++. Not all jobs are well suited for C, but I like it, so I tend to use it even if I might be better off with a different choice.

Java—a language developed by Sun Microsystems—is not unlike C++ (which is the predominant object-oriented extension to C). The core language shares a great deal in common with C++ even down to most of the syntax. There are two major differences:

1. Java is strictly an object-oriented system. You can use C++ without using objects, but Java requires you to use objects at all times.

2. The Java library is particularly well suited for the Internet.

If you don't know object-oriented programming, don't worry. It requires you to change how you think a little, but the payoff is well worth the effort. If you've programmed in virtually any other language, you'll find Java is simple to learn. If you've looked at books about Java before, you may have been put off by the complexity of the example programs. That's because most books concentrate on graphical user interfaces, which are complex by their very nature. In an embedded system, programs are usually much more straightforward.

This book won't teach you every detail of Java, but it will give you enough background that you'll have no trouble doing the type of Java you'll use in a typical embedded system. At the end of the chapter, you'll find some pointers to places to continue your Java education. You'll find the exercises and experiments in this chapter will give you a great head start at becoming a dyed-in-the-wool Java programmer.

Experimenting with Java

One of the nice things about Java is that it works on nearly any computer system. This chapter will show you the fundamentals of Java, and you can try the example programs on practically any computer you have—a PC, a Mac, or a Linux computer will do just fine.

You will need a Java development environment. If you already have a tool that you are comfortable with, you may use that. Otherwise, you can download a free Java Developer's Kit (JDK) from java.sun.com. When you download the JDK, be sure to download the correct version for the operating system you use.

The JDK's tools are command-line-oriented. You'll use javac to compile your programs into *.CLASS* files. The Java program can execute the program. If you just installed the JDK, you might want to try the program in Listing 2-1 just to see that everything is working properly. If you type the code in by hand, be aware that Java is case-sensitive.

With Java, the file name is important, so the code in Listing 2-1 must be in a file named *Test.java*. That's because the first line identifies the public class in this file as having the name *Test*.

To compile this program using the JDK, you'll enter the following command line:

Listing 2-1
A Very Simple Java
Program

```
public class Test
{
 public static void main(String args[])
  {
  System.out.println("Hello!");
  }
}
```

```
javac Test.java
```

Then when you want to run the program, enter

```
java Test
```

You should see the output string on your screen.

Java: Compiled or Interpreted?

Obviously, the javac program is a compiler. However, it is not a compiler in the same sense that is, say, a traditional C compiler. A classic compiler transforms source code into machine code that the computer can execute. Other languages, like Basic, are often interpreted. An *interpreter* is a program that examines source code and performs the actions that it specifies.

Java is a mix of these two techniques. The javac program converts your source code into a make-believe machine code designed to run on a Java Virtual Machine (JVM) and stores the result in class files. You can think of the JVM as a fictional microprocessor and class files as the executables that run on it. Of course, your computer doesn't have this microprocessor, so the Java program simulates the JVM and interprets the class files. Therefore, your Java program is both compiled and interpreted. A few JVMs actually further compile the class files on demand to improve performance.

In the embedded system world, you often won't have room for the entire JVM in your target processor. In that case, you may have to precompile your program into another format specifically for the target. You'll see specific examples of this in later chapters.

If you are accustomed to languages like C++, you might wonder where the link step is. *Linking* is the process of bringing together separate parts of your program, libraries, and necessary runtime elements into a single executable file. Java performs these functions automatically at runtime. Your Java program can load any class file it needs at any time. That's how applets work in Web browsers. The Web browser can't know ahead of time what applet a Web page might want to load. With Java's runtime linking, that's not a problem. The browser simply copies the applet's class file over the network and then links to it. That's why a class file that contains a public class named *Test* must be *Test.class*. Without this restriction, the JVM would have to search every class file when you wanted to load an object. Of course, that means a class file can only contain one public object—that is, objects accessible to any other program. It is possible to have objects that are not public, and there are fewer restrictions on these classes.

Getting Started

Every Java program consists of at least one public class. Of course, larger programs may consist of many classes of different types. If you expect to run the program from the command line, the class must contain a static *main* function similar to the one in Listing 2-1. This is the entry point function, and you can access any command line arguments using the *args* array passed to *main*.

Soon you'll learn more about objects, but for now you can simply consider the code in Listing 2-1 as boilerplate. Replace *Test* with your desired class name, and put your program in the *main* function (replacing the call to *System.out.println*).

You probably noticed that all statements end with semicolons. This is similar to C or C++. You usually won't use a semicolon before an open brace ({), because the braces enclose a group of statements that the compiler treats as a single statement.

Variables, Types, and Constants

Consider the program in Listing 2-2. This program uses a single variable, *i*, that stores values. The variable is an *int*, and so it will hold an integer value. Java has only a handful of fundamental types like *int* (see Table 2-1). On an embedded microcontroller, you may not have all of these types available. For example, some processors won't support any floating point numbers. You'll often make variables that are of a specific object type (like *String*). You'll learn more about object variables later.

In Listing 2-2 you can see the variable declaration (*int i*) and an assignment statement that computes a value and stores the result in *i*. Names are case-sensitive in Java, so it is possible (although not a good idea) to have another variable named *I*. You can assign a value when you declare a variable, as in this example:

Listing 2-2
Counting in Java

```
public class Test2
{
int usecount;
public static void main(String args[])
  {
    int i;
    i=33*9;
    System.out.println(i);
  }
}
```

TABLE 2-1

Fundamental Data
Types

Type	Description
boolean	True/false value
char	Unicode character
byte	8-bit signed integer
short	16-bit signed integer
int	32-bit signed integer
long	64-bit signed integer
float	32-bit floating point number (IEEE 754-1985 format)
double	64-bit floating point number (IEEE 754-1985 format)

```
int i=10;
```

You can also define multiple variables of the same type:

```
int i,j,k=33,loopctr=0;
```

You'll notice that the variable declaration is within the *main* function. That means that it is local to the function. If there were more than one function, the *i* variable would not be accessible to the other functions. There is really no such thing as a global variable in Java, but you can come close by using fields. A field is a variable that belongs to an object. For example, *usecount* in Listing 2-2 is a field. You'll read more about fields later in this chapter.

Constants

Sometimes you'd like to make a variable that has a constant value. For example, you might want to write

```
double pi = 3.14159;
```

However, there is no reason for your program to ever change the value. It is a constant. The way it is written, your program could—perhaps by accident—change the value. Also, the Java compiler has no way to know that the value will never change, and it might be able to generate better code if it knew that was the case.

To solve this problem, you can make the variable *final* (or *static final*—a distinction you'll read about shortly). This means that it can't be changed and the compiler knows it. That also means you should initialize the variable, since you can't change it later. For example:

```
static final double pi = 3.14159;
```

One thing to watch out for is that Java always treats ordinary literal numbers as either *int* or *double* data types. So consider this line of code:

```
static final float pi = 3.14159;
```

This actually causes a type conversion because 3.14159 is of type *double*—the compiler will complain. You could convert the value to a *float,* but it is easier to specify the constant as a *float* to begin with by adding an *f* suffix:

```
static final float pi = 3.14159f;
```

If an integer is wide enough to fit in a *short* or a *byte,* the compiler will automatically handle the conversion. However, if you want a *long* literal, you must suffix it with *L* (or *l,* but the uppercase L is better because it is easy to confuse a lowercase L with a numeral 1).

You can create literal characters by using single quotes around any Unicode character. So:

```
char stop='X';
```

Table 2-2 shows some escape sequences used to generate special characters (like a single quote, or a new line). You can also use a C-style escape,

\ddd (where *ddd* is the octal value of a character) or the Unicode escape *\uXXXX* (where *XXXX* is the hex value of the character). String literals follow the same rules, but you enclose them in double quotes, not single quotes.

Number Bases

You can also specify literal integers in octal (base 8) or hexadecimal (base 16). Octal numbers have a 0 prefix, while hexadecimal (or hex) numbers have a 0x prefix. This can be tricky. Consider this code fragment:

```
int x=010;
System.out.println(x);
```

The result printed is 8, because the leading zero marks the literal 010 as an octal number.

Expressions

When you write x=10+3, x=x+1, or even x=0, you are assigning an expression to the x variable. Expressions combine variables and constants using operators (see Table 2-3). Consider this problem:

```
x=5+3*2
```

TABLE 2-2	Sequence	Meaning
Escape Sequences	\n	New line
	\t	Tab
	\b	Backspace
	\r	Carriage return
	\f	Form feed
	\\	Backslash
	\'	Single quote
	\"	Double quote
	\xxx	Any character (xxx is octal number)

TABLE 2-3

Basic Java
Operators

Operator	Definition	Operator	Definition
++	Post- or preincrement	<	Less than
--	Post- or predecrement	<=	Less than equal to
~	Bitwise invert	>	Greater than
!	Boolean invert	>=	Greater than equal to
*	Multiply	==	Equal to
/	Divide	!=	Not equal to
%	Remainder from integer division	&	Bitwise AND
+	Addition	∧	Bitwise exclusive OR
−	Subtraction	|	Bitwise OR
<<	Left shift	&&	Logical AND
>>	Right shift with sign extension	||	Logical OR
>>>	Right shift	?:	Conditional

The answer depends on which you do first, the addition or the multiplication. If you add first, the answer is 16. If you multiply first, the result is 11. The correct answer—by convention—is 11 and that is what Java will return. Table 2-4 shows the precedence of the various operators. Operators that appear first in this table occur before ones that appear lower. In the above example, the * operator is on line 4 and the + operator is on line 5, so the result is 11. You can override the order by using parentheses. If you wanted the answer to be 16, you could write

```
x=(5+3)*2
```

Given equal precedence, Java evaluates left to right. So 4 + 2 + 9 is the same as (4 + 2) + 9. This is usually not important, but it can be a factor in some complex expressions.

Special Operators

For the most part, the Java operators will be familiar to you if you've used any other programming language. A few, however, may seem odd if you haven't used C or C++ before.

TABLE 2-4

Order of
Operations

[]. (params) expr++ expr --	Highest priority
++expr -expr +expr -expr ~!	
new (typecast)	
* / %	
+ -	
<< >> >>>	
< > >= <= instanceof	
== !=	
&	
∧	
&&	
‖	
?:	
= += -= *= /= %= >>= <<= >>>= &= ∧= |=	

For example, the ++ and -- operators can be confusing. These special operators increment or decrement (that is, increase by one or decrease by one) the variable they alter. Instead of writing

```
foo = foo + 1
```

You might write

```
foo++;
```

That doesn't seem like a big improvement, but you can also use these operators within other expressions. If the ++ occurs before the variable, the increment occurs before Java uses the value. If it occurs after the variable, the increment occurs after Java uses the value. You'll understand how this works if you consider the following code:

```
int x=10;
int y=3*++x;   // y = 33 and x=11
int z=2*x++;   // z = 22 and x=12
```

If you want to increase the value by more than just one, you can write:

```
x=x+10;
```

This works with +, −, /, and ` as well.

Another operator that is unusual is the conditional operator. This operator requires three arguments. The first is a Boolean. If the Boolean value is *true*, the operator returns the second expression. Otherwise, it returns the third. For example, this statement assigns x with 0 if y is 10, or 100 in other cases:

```
x=y==10?0:100;
```

Notice that two equal signs are the operator that tests for equality. A single equal sign is for assignment only.

You might wonder about the difference between the & and && operators (or the | and ∥ operators). The single-character operators do bitwise operations. In other words, & takes the bits of its two arguments and *and*s them together. The double-character versions only work on Boolean values.

Comments

It is always a good idea to add comments to your code. This helps other people understand your program and might even help you figure out what you were doing when you return to your code a few weeks or months after you wrote it.

Java allows you to start a comment with two slash characters (//). After the two slashes, Java ignores everything else on that line. If you want to make multiline comments, start them with /` and end them with `/. However, /`` is a special type of comment known as a *Java Doc comment*. A special program (javadoc) can scan Java source code and use special commands embedded in Java Doc comments to automatically create documentation in HTML or other formats.

Control Flow

All programming languages need a way to control a program's flow. Otherwise, your programs would be just a list of commands.

What's Unicode?

Most people are familiar with ASCII. That's a standard mapping between 7-bit numbers and characters. For example, 65 is an uppercase A and 32 is a blank. There are 128 possible ASCII characters. Of course, most computers store ASCII as bytes, and so you can really store numbers from 0 to 255, which gives you 128 extra characters.

There are two problems with this. First, there is little agreement about what those extra characters are. While there are a few standard conventions, none are as widespread as the ASCII standard. Besides that, even 255 characters are not really enough to handle the numerous characters required if you plan on making your programs handle multiple languages. Sure, the usual character set handles English, but many languages use characters with special accent or other marks. Then there are languages that use Cyrillic or other characters that ASCII is not able to handle.

To address this problem, many programs are turning to Unicode. Developed by the Unicode Consortium, Unicode is a standard that provides 16-bit characters. With over 65,000 characters available, Unicode has many more characters than the ASCII set, including many graphical characters and letters to support many languages other than English.

All strings in Java are represented by Unicode. Of course, your computer's input and output system might use ASCII, so Java can translate between the two character sets. You can convert any ASCII character to Unicode (in fact, the first 128 Unicode characters are identical to the ASCII characters). However, not all Unicode characters have ASCII equivalents, as you might expect.

To solve this problem, Java can utilize UTF/8 (Unicode Transmission Format/8; defined in RFC2044) to store proper Unicode characters on a byte-oriented device. The idea behind UTF/8 is simple. Since each Unicode character from 0 to 127 is an ASCII character, UTF/8 represents these as their ASCII equivalents. That means a pure ASCII UTF/8 file and a pure ASCII file are identical.

If the UTF/8 file requires Unicode characters beyond 127 (7F hex, or 0x7F), UTF uses a special marker bit pattern and a varying number of bytes to identify the character. For characters ranging between 0x80 and 0x7FF, UTF/8 uses two bytes. The first byte will have its first 2 bits set to 1, and the third bit set to 0. The remaining 5 bits will be the most significant bits of the character. The second byte will have its first bit set to 1 and the second bit set to 0. The remaining 6 bits will have the least significant bits of the character.

That scheme handles numbers up to 11 bits long. For characters 0x800 and beyond, UTF/8 stores the first byte with the binary pattern 1110xxxx (where the x's are the most significant part of the character). The next two bytes are in the form 10xxxxxx (the same format as the second byte in the 11-bit scheme).

This has several important implications:

- Any program that only deals with straight ASCII characters will never see any difference.
- If anything in the file looks like an ASCII character, then it is. None of the extra bytes appear to be ASCII because they are all greater than 0x7F.
- Any byte that starts with binary 10 is the middle of an extended character. The first byte of a character always begins with a binary 0, or a binary 11.

Java uses Unicode fully. For example, in this chapter you've read about making a constant like this:

```
static final float pi = 3.14159f;
```

However, if you know how to make the correct Unicode character with your keyboard, you might just as well write:

```
static final float π= 3.14159f;
```

In practice, you rarely see this since most users still use ASCII-oriented keyboards. Java does allow you to specify any Unicode character by using \uXXXX (where *XXXX* is the hex number that represents the character).

Java supports the traditional *if* statement along with *for, while,* and *do.* These work nearly the same as their C counterparts. Listing 2-3 shows a simple program that uses a *for* loop. The first expression in the *for* statement sets the initial conditions. The second expression tests for the end of the loop, and the final expression modifies the loop variable at each loop.

Even if you are used to C or C++, Java's strong typing can throw you a few curves. For example, in C++ you might write

```
int t;
t = somefunction();
if (t)
   dosomething();  // call if t is not zero
else
   dosomethingelse(); // call if t is zero
```

This won't work in Java. Why not? The variable *t* is an integer but the *if* statement expects a Boolean value. You'd have to write

Listing 2-3
Simple Program
Using a *for* Loop

```
public class forDemo
{
public static void main(String[] args)
 {
 int i;
 for (i=0;i<10;i++) System.out.println(i);
  }
}
```

```
int t;
t = somefunction();
if (t==0)
   dosomething();  // call if t is not zero
else
   dosomethingelse(); // call if t is zero
```

Another place where Java differs from C is in the *break* and *continue* statements. In Java, as in C, you use these statements to either end a loop (in the case of *break*) or go directly to the next iteration of the loop (for *continue*). However, these statements have extra features in Java.

Consider this loop:

```
for (i=0;i<10;i++)
   {
   System.out.println(i);
   if (func(i)==3) break;
   if (i%2==0) continue;  // don't do any more for even numbers
   System.out.println("Odd number");
   }
```

The *break* statement, if executed, immediately terminates the loop. The *continue* statement just moves on to the next iteration of the loop (in this case, that prevents even numbers from getting to the bottom of the loop).

Unlike C, Java allows you to include a label as the target of a *break* or *continue*. This lets you terminate or continue nested loops. For example:

```
Loop0:
for (x=1;x<10;x++)
   {
   for(y=1;y<20;y++)
      {
      .
      .
      .
      if (checkexit()==true) break Loop0;
```

```
    }
}
```

The *for* loop above, by the way, is functionally the same as this code:

```
int i=0;
while (i<10)
  {
  .
  .
  .
  i++;
  }
```

There will be many times when you want to test a value against several constants and take particular actions depending on the value. You could write a series of *if* statements, of course. However, Java provides the *switch* statement, which is more succinct. Listing 2-4 shows an

Listing 2-4
Example of Java
Using *switch*

```
public class swDemo
  {
  public static void main(String[] args)
    {
    int x=Integer.parseInt(args[0]);
    switch (x)
      {
      case 1:
        System.out.println("Number one");
        break;
      case 2:
      case 3:
        System.out.println("Either 2 or 3");
      case 4:
        System.out.println("Either 2, 3, or 4");
        break;
// This is a catch-all case
      default:
        System.out.println("You didn't enter 1-4!");
      }
    }
  }
```

example of using *switch*. Notice that once a match occurs, the code executes from that point—even if it encounters another *case* statement. This allows you to cascade several cases that share the same code. However, most often you'll want each case to be separate and you'll want to write a *break* statement at the end of each *case*.

Classes and Objects

The most important feature of Java is the *class*. A class is nothing more than a definition of data items—known as fields—grouped together. Along with the fields, the class also has functions (known as *members*) that operate on the data.

A class is to an object as a cookie cutter is to a cookie. A class is a schematic that allows you to make objects. For example, consider a class that represents a thermostat used in a building's air conditioning system. The class might have fields to represent the current set point temperature and the current actual temperature. In addition, there might be methods that request an update of the current temperature or a manual override to turn the system on and off. You can see an excerpt of this imaginary class in Listing 2-5.

Listing 2-5
Example of a Class for the Thermostat of a Building's Air Conditioning System

```
Class Thermostat
{
private int id;
private float setpoint=20.0f;
public Thermostat(int _id) { id=_id; }
public void setTemp(float temp)
  {
   . . .
  }
public float getTemp()
  {
   .
   .
   .
  }
```

All by itself, this class does nothing. If you want to represent a particular thermostat, you'll have to instantiate the object. First, you'll declare a variable of the object's type:

```
Thermostat t1;
```

This isn't an object; it is simply a reference to an object. What's the difference? For one thing, *t1* does not contain a *Thermostat* object—it simply refers to an object that exists on its own. To initialize *t1*, you'd write

```
t1=new Thermostat;
```

This creates a *Thermostat* object and, coincidentally, assigns a reference to the new object to *t1*. You might also write

```
Thermostat t2=t1;
```

Now *t2* and *t1* refer to the exact same object. That means you can change the object using either *t1* or *t2* and it will have the same effect.

This has an odd effect when testing for equality. If you test to see if *t1* and *t2* are equal (using ==), the result will be true if and only if the two references point to the same object. For thermostats, that is probably the right thing to do. But consider objects like *String* (the built-in object for handling text strings). Try the code in Listing 2-6. It reports that the two strings are not equal when they clearly are the same. That's because using == tests to see if the variables refer to the exact same object, and

Listing 2-6
Testing for Equality

```
public class Test
  {
  public static void main(String args[])
    {
    String s1="Al";
    String s2="A";
    s2=s2+"l";
    if (s1!=s2) System.out.println("Not");
    System.out.println("Equal");
    }
  }
```

s1 and *s2* do not. Many objects (including *String*) provide an *equals* method that tests for logical equivalence. If you change the test to *s1.equals(s2)*, you'll find the program works as you'd expect.

Methods and Parameters

The *equals* method is a common method that exists in every class. Of course, you can write your own methods (or functions, if you prefer). Each method belongs to a class and returns a value. Methods can also take arguments or parameters. You can have two methods in the same class that have the same name as long as they accept different parameters. For example, you might have a method known as *print* that accepts an integer argument and another one that accepts a *String*. From Java's point of view, these are two entirely different functions.

Methods return values (using the *return* statement). If you don't need to return anything, you can define the method as a *void* type. If you don't specify *void*, then you must use a *return* statement or you'll get a compile error.

Classes can contain special methods that have the same name as the class. These special methods are constructors and have no return type. They can, however, accept arguments. You can have multiple constructors with different argument lists.

Consider the simple class in Listing 2-7. Here the *construct* object has three fields. The *intval* field can store an integer value and the *strval* field stores a string. The *which* field tells which of the two values was set (if any). Notice there are three constructors. One takes no arguments (the default constructor). The other two take arguments of the appropriate type. Each constructor sets the correct field and the *which* field as appropriate.

When you use *new* to create a new instance of an object, you can provide arguments, as in

```
c1 = new construct(10);
```

One of the most prominent features of Java is that it automatically destroys objects when they are no longer in use. This is known as *garbage collection*. The way it works is that Java keeps track of references to objects. After you are done with an object, Java will eventually destroy it and reclaim its resources.

Usually, this happens without any explicit action on your part. For example, suppose you create an object within a method call. When the

Listing 2-7
Simple Class

```java
public class construct
  {
  final int NONE=0;
  final int INTEGER=1;
  final int STRING=2;
  int intval;
  String strval;
  int which;
  public construct()
    {
    which=NONE;  // no value set
    }
  public construct(int value)
    {
     intval=value;
     which=INTEGER;
     }
  public construct(String value)
    {
     strval=value;
     which=STRING;
     }
  }
```

method returns, the object reference goes out of scope. If there are no other references to the object, Java will, at some point, reclaim the object's resources. This might happen right away, or it may happen much later, when the Java system is low on memory.

For local variables in a method, that works well. However, if you have static variables or object fields that have a long life, you may want to release an object early. You can do this by assigning *null* to the object reference. Once there are no more references, the object is subject to garbage collection. Another problem that this can solve is when objects refer to each other. Suppose object A refers to object B and object B refers to object A. Even when you are no longer using either object, Java can't garbage collect these objects because they still have a reference active—a reference to each other! Assigning a *null* into the references before you finish with the objects can solve this dilemma.

In C++, you destroy objects explicitly, and you can write a special destructor function that handles the cleanup. In Java, you don't really know when the object will vanish, but you can write a *finalize* method that Java will call before it garbage collects an object.

Where Are the Pointers?

Speaking of C++ and *null,* if you are familiar with C++ or assembler language you might wonder how Java handles pointers. A common misconception is that Java doesn't have pointers. This is not really true. In Java, every time you use an object you are using a pointer to the object. That's why you say an object variable is a reference, not the object itself.

For example, suppose you want to create a linked list. Each item in the list has a reference to the next element. Listing 2-8 shows a simple class that implements the elements. The test *main* method builds a simple list with four elements. Notice that the program has only one variable that holds a reference to a list element (*head*). However, the other objects won't be garbage-collected, because each object holds a reference to the next one.

Every object has a special pseudoreference known as *this.* You can use this to refer to the current object. You can see this in Listing 2-8 where the *List* object's *insert* method sets the *next* link.

There are a few more interesting points to Listing 2-8. First, notice that *head* is static. There is only one head reference no matter how many list items are in use. What's more, the *printList* method is also static. This is for the same reason—it applies to the list as a whole. The *for* statements that scan the list are a good example of using a *for* loop in a nonnumeric situation. Remember, the first clause initializes the loop (*ptr=head*). The second clause tests for the end condition (*ptr==null*), and the third clause sets up the next iteration of the loop (*ptr=ptr.next*). These clauses are not the usual numeric expressions, but they still work.

In the test *main* program, you'll see four *new* statements that create objects. They look a bit peculiar because the program doesn't store the object reference anywhere. Instead, it simply calls *insert* directly. Because the list maintains the references, the objects will not be subject to garbage collection. Since the program no longer needs the objects, there is no need to retain a reference to them. To print the list, the program uses the *printList* method.

Listing 2-8
Simple Class That
Implements the
Elements

```java
public class List
{
static List head=null;  // pointer to first item
String value;
List next;
// create list element (not linked)
List(String s)
  {
  value=s;
  next=null;
  }
// insert item in list
void insert()
  {
  List ptr, last;
  if (head==null)
    {
    head=this;
    return;
    }
// this code finds the last item in list
  last=head;
  for (ptr=head;ptr!=null;ptr=ptr.next)
    last=ptr;
  last.next=this;
  }
  static void printList()
   {
   List ptr;
   for (ptr=head;ptr!=null;ptr=ptr.next)
    System.out.println(ptr.value);
   }
  static public void main(String args[])
   {
   new List("One").insert();
   new List("Two").insert();
   new List("Three").insert();
   new List("Four").insert();
   List.printList();
   }
  }
```

Arrays

Java also supports array data types. You can create arrays of basic types (like *int*) or you can create arrays that contain object references. All arrays in Java are actually objects. You create them using a syntax similar to an object:

```
int [] x;   // reference to array
x = new int[33];  // create array with 33 elements
```

You can also use an alternate syntax to declare the array reference:

```
int x[];
```

Given the above declaration and *new* statement, you could refer to the first element of the *x* array as *x[0]*. The last element is *x[32]*. You can use these just like any other variable:

```
x[2]=17;
system.out.println(x[2]);
```

Since arrays are really objects, they may have fields. The one you'll find particularly useful is the *length* field. This allows you to determine how many elements the array contains. This is very useful when you want to loop through the entire array with a *for* loop (see Listing 2-9).

Notice in Listing 2-9 that *testary2* uses a set of constants enclosed in brackets as an initializer. This is known as an *array literal,* and it only works when declaring an array variable. In version 1.1 and later of Java, you can use a similar syntax to create a nameless array anywhere in your code. Consider this method call:

```
selectColor(new String[] { "Red","Green","Blue" });
```

Here the code creates a new initialized array of *String* objects.

Strings

You usually don't think of strings as relating to embedded systems, but these days many embedded systems do manipulate strings. You might want to write to an LCD, or receive commands from a PC or to a modem. What's more, Internet programs constantly format and receive strings as they communicate with other computers on the network.

Listing 2-9
Looking Through an
Array

```
public class ary
  {
  public static void main(String [] args)
    {
    String [] testary;
    String [] testary2 = { "One", "Two", "Three" };
    testary=new String[5];
    int i;
// initialize testary
    for (i=0;i<testary.length;i++)
      testary[i]=Integer.toString(i*2);
// print both arrays
    for (i=0;i<testary.length;i++)
      System.out.println(testary[i]);
    for (i=0;i<testary2.length;i++)
      System.out.println(testary2[i]);
    }
  }
```

You saw earlier in the chapter that Java doesn't use ASCII internally. It uses Unicode—a 16-bit character standard. Java's *String* class, of course, stores characters as Unicode as well.

Strings are objects, but they are so prevalent in many programs that Java makes a special concession to them. You can still create *String* objects using *new* like any other object. You can also assign a string literal to a *String*. For example:

```
String modemprefix = "AT";
```

Like all objects, *String* objects have fields and methods. If you are C programmer, you might think of *String* as similar to an array. However, in Java, strings have very little in common with arrays.

One surprising feature of *String* is that, once set, the actual *String* object never changes. That's not to say that the reference can't change, but the actual object stays the same. This can lead to performance problems if you are not careful. For example, suppose you have a method named *getC* that retrieves a character from some source. You might write this code to build a *String* object in the *s* variable:

```
String s = new String();
for (i=0;i<1000;i++) s=s+getC();
```

This will work, but it is very inefficient. When you compute $s + getC()$, you create another *String* object. Then you set the *String* reference *s* to point to that new object. That means the original string now has no references, and will be subject to eventual garbage collection. Throughout this loop you'll create and discard 1000 *String* objects! Remember, Java doesn't actually garbage-collect on each loop, but eventually the JVM will recycle the objects.

To prevent this problem, Java also provides a *StringBuffer* object. These objects are similar to *String* objects, but they allow you to manipulate characters in place. Once you are done, you can convert the *StringBuffer* into a proper *String*:

```
StringBuffer sb = new StringBuffer(1000);
String s;
for (i=0;i<1000;i++) sb.append(getC());
s=sb.toString();
```

The *String* object has several useful methods (see Table 2-5). Most of these are straightforward, although many people have trouble with *substring*. The *substring* method has two versions. One takes the starting index and returns the substring from that index to the end of the string. The other version takes a starting index and an ending index. This version returns the string starting at the first index, and ending at the character *before* the second index. Consider a string that contains "WD5GNR." The index arguments start at 0, so if you call *substring* with arguments of 2 and 4, the call will return "5G," not "5GN" as you might expect.

Extending Classes

The biggest benefit to object-oriented programming is the ease with which you can reuse code. One thing that makes this possible is *inheritance*. The idea behind inheritance is that each class extends another class and inherits methods and fields from this base class. Suppose you have a class that represents a temperature probe (see Listing 2-10).

Later, you update the sensor to include a wind speed indicator. Instead of maintaining two copies of the temperature code, you can create a new class (*DeluxSensor*; see Listing 2-11) that extends the temperature sensor code. In this way, all the code and fields in the original code are available in the new class. If you make changes to the original code, the

TABLE 2-5	Method	Description
Useful String Methods	valueOf	Convert basic types (*float, int, boolean*) to a *String*
	charAt	Retrieve one character from a *String*
	compareTo	Compare one string to another
	compareToIgnoreCase	Compare strings without regard to case
	endsWith	Test for suffix
	indexOf	Search for character or substring within string
	lastIndexOf	Reverse search for character or substring within string
	replace	Substitute one character for another
	startsWith	Test for prefix
	substring	Returns portion of string
	toLowerCase	Convert to lowercase
	toUpperCase	Convert to uppercase
	trim	Remove leading and trailing white space

Listing 2-10
Class Representing a
Temperature Probe

```
public class Probe
{
public Probe(int portnum) { . . . }
public float getTemp() { . . . }
public void setOptions(int a) { . . . }
}
```

new object will inherit the same changes automatically. In this case, the original sensor object is the base class. The new object is said to extend (or derive from) the base class.

It is possible to extend this hierarchy to any number of levels. For example, you might extend *DeluxSensor* into *WeatherStation* that integrates several instruments and an LCD interface. However, unlike some languages, Java only allows you to derive from a single class—it is not possible to derive directly from more than one class.

If you don't specify a base class, your class will extend the default *Object* class. That means that all objects, no matter what their type, will

```
public class DeluxSensor extends Probe
{
public DeluxSensor(int portnum) { . . . }
public float getWindSpeed() { . . . }
public float getWindDir() { . . . }
// getTemp and setOptions are inherited from Probe
}
```

have the basic methods that belong to *Object* (see Table 2-6). Remember, classes that extend other classes (including *Object*) can (and often do) replace methods with custom versions. For example, quite a few classes override *toString* to provide a more meaningful string representation of their contents (the default *toString* doesn't print any of the object's contents). Many objects (like *String*) override *equals* to test the object's contents instead of the actual object.

Usually, you'll want to allow others to extend your classes and inherit members (that is, methods and fields). However, you can control what other classes can access. If you name certain fields or methods *private*, they will not be accessible by code in any other class (including classes that extend this class). If you mark members *public*, any code can access them. You can also specify members as *protected*. Classes that extend your class can freely access *protected* members, but other classes have no access. If you don't specify any of these access modifiers (that is, *private, public*, or *protected*), the member is accessible to any code in the same package. You'll read more about packages shortly, but for now consider a *package* as one subdirectory. In addition to making certain members private, you can also mark a class as *final*. This will prevent other classes from extending your class.

Just because a base class provides members doesn't mean the derived class has to use them. You can override functions (or fields) when you want to provide replacements. You can still call the base class version by using the *super* keyword. This can be useful if you want to make a minor modification to an object. For example, suppose your temperature sensor class operates using Fahrenheit temperatures. Later, you decide you want to create a version to do Celsius temperatures. You can simply extend the original class and override the *getTemp* function. Instead of rewriting it totally, you can still call the original class function:

```
float getTemp()
{
```

TABLE 2-6	Method	Description
Object Methods	equals	Tests objects for equality
	getClass	Returns class object that created this object
	hashCode	Returns id number (hash) for this object
	notify	Used by other threads to notify threads waiting on this object that conditions changed (releases one waiting thread)
	notifyAll	Same as notify, but releases all threads waiting
	toString	Returns a string representation of the object
	wait	Causes this thread to wait until another thread calls this object's *notify* or *notifyAll* methods
	clone	Duplicates object
	finalize	Called before garbage collection

```
return 5*(super.getTemp()-32)/9;
}
```

This is a common theme in embedded programming. For example, you might have a base class that represents a serial port. You could extend the class to represent instruments that use the serial port. That way all the serial port code resides in the main class and the other derived classes can share that common code.

An important consequence of using derived classes is *polymorphism*. Polymorphism is a simple concept for such a fancy word. Suppose you've built the serial port class and extended three other classes from it: *Temp, Wind,* and *Humid.* These classes—of course—represent different instruments that all use a serial port for communications. What if you want to keep a list of these items in an array? Since they are all derived from *SerialPort,* you can treat them as if they are *SerialPort* objects. Listing 2-12 shows how it works. Once you place the objects in the *instruments* array, you can't use members that belong to the derived classes. In other words, calling *instruments[0].getTemp()* is not legal. However, you can access anything that belongs to *SerialPort.* For instance, if *SerialPort* defines an *init* method, you could call it using any (or all) of the elements of the array. If any of the specific objects override the *init* function, Java will call the correct override.

This is not true, however, of fields. If the *SerialPort* object defines a field named *port* and *Humid* overrides it, you'll access different fields

```
public class Instruments
{
public SerialPort[] instruments = new SerialPort[3];
public Instruments()
{
instruments[0] =new Temp(1);   // on port 1
instruments[1] =new Wind(2);
instruments[2] =new Humid(3);
}
```

depending on if you are using a *SerialPort* variable or a *Humid* variable. That's true even if the *SerialPort* variable really refers to a *Humid* object. Remember, variables are just references to objects, and it is legal for a base class variable to refer to a derived class object.

If you want to force an object reference into another type of object, you can use a *cast*, which is simply the name of the object in parentheses. You can only cast an object to a correctly related class. For example, you can cast any object to *Object* since it is a base class of all objects. You can also cast an object back to its original class. However, you can't cast an object to a class that doesn't appear in the object's class hierarchy.

Suppose you have class B that extends class A. You also have class C that doesn't extend any other class (except, of course, *Object*, which is the default). Further suppose that you have the following declarations:

```
B b = new A();
B b1;
A a;
C c = new C();
```

The following assignments are legal:

```
a=(A) b;
b1=(B)a;
```

However, the following is not legal:

```
a=(A)c;
```

That is because class A and class C are not related. You also could not make the following assignment:

```
b1=(B)new A();
```

Constructors present a special problem. Each class has to provide its own constructors. Then, if you don't do anything special, Java calls the default constructor for each base class, starting with *Object* (which is the ultimate base class of every object) and working down the class hierarchy until, finally, the most specific constructor executes.

If you think about this, it makes sense. After all, a derived class might need to use functions in the base class that require that the base class's constructor has already executed. However, there are a few cases where this chaining of constructors doesn't work correctly. For example, suppose the base class doesn't have a default constructor? The same situation might arise when the derived class needs to call a nondefault constructor.

The answer is to make the first line of the derived constructor a call to *super.* This special keyword calls the base class constructor explicitly. See Listing 2-13 for an example.

Basic Type Classes

Nearly every data type you can use in Java is an object. Since all objects derive from *Object,* that means you can depend on a certain number of methods being available in all objects. For example, *toString,* a function

Listing 2-13
Deriving One Class
from Another

```java
class baseclass
   {
   private int val;
   public baseclass(int x) { val=x; }
   public int getVal() { return val; }
   }
class extender extends baseclass
   {
   private int val2;
   public extender(int a, int b)
      {
      super(a);
      val2=b;
      }
   public int getAltVal() { return val2; }
   }
```

in *Object,* returns a string representation of any object. Many objects override *toString* so they can return a meaningful representation.

What about the basic types like *int* and *float?* Often, it is useful to have a class that represents one of these types. However, you don't want the overhead of using an object just to perform simple operations. Therefore, Java uses simple types for most purposes, but also provides corresponding objects. For example, the *Integer* class wraps an *int* value, and the *Float* class wraps a *float* value.

This has several benefits. First, you might want to treat a basic type as an object so you can put it in an object array with other objects. Also, these objects act as a central clearinghouse for functions related to the type. Remember, Java has no real global variables or methods—everything has to belong to a class.

Numeric Conversions

You'll often use the wrapper classes to convert strings to the appropriate type. For example, *Integer* has two functions (*parseInt* and *valueOf*) that convert strings to integers. The *parseInt* method returns an *int,* whereas the *valueOf* function returns an *Integer* object. You can also specify an optional radix if you want, for example, hex or octal interpretations.

In the opposite direction, you can use *toString* to convert an integer to a string. You can also use functions like *floatValue* or *longValue* to convert an *Integer* object to a different type. To convert the basic types, you can use a cast:

```
int n=100;
float fn = (float) n;
```

Statics

Numeric conversions show one of the uses of the wrapper classes—Java uses them as containers for what might otherwise be global functions. It does this using static methods. This allows you to refer to a function without having to actually create an instance of an object. Suppose you have an integer variable *x.* You can't call *toString* on an *int* because it isn't an object. You could construct an *Integer* object to contain the *int,* but that's a lot of work just to do a string conversion.

Luckily, *Integer* provides *toString* as a static member, so you can call it like this:

```
String s = Integer.toString(x);
```

You can make methods or fields *static.* Be aware that a static method can't access any normal fields or methods directly, because there is no object instance associated with the static method. Therefore, there is no *this* reference. That also means, in the case of fields, that there is only one copy of the variable no matter how many object instances exist. That makes static fields useful for creating a kind of global variable. If you make the field *public,* any other part of your program can access the variable (using the class name as a prefix). If you make the field *private* or *protected,* the variable will still be like a global variable, but it won't be accessible from other objects (or unrelated objects in the case of *protected*).

Interfaces and Abstraction

One of the problems with Java's class extension is that one class can only extend one other class. Sometimes it would be useful to extend multiple classes. For example, suppose you had a serial port class and a thermostat class. You might want to create a new class to represent a thermostat that uses a serial port. Some languages allow you to use more than one class as a base class, but Java is not one of them.

To solve this problem, Java introduces the notion of *interfaces.* An interface is similar to a class, but it doesn't contain any program code. Instead, it defines methods that other classes must implement.

Suppose you have a system that has a variety of objects. You'd like to print reports using these objects, and each object should contain the necessary code. However, the objects can't all inherit a common base class. You could write a method, say *report,* in each class. However, this would make it difficult to write a single piece of code that could generate the report using a variety of objects.

The answer is to use an interface. Suppose you write this interface definition:

```
Interface Reporter
{
void report(OutputStream out);
}
```

This interface defines a method, but doesn't provide any code for it. Interfaces never include code, only definitions. Now you can define classes using this interface:

```
class ClassA extends baseclass1 implements Reporter
{
.
.
.
void report(OutputStream out)
 {
 .
 .
 .
 }
}
```

Because the class declaration contains the phrase *implements Reporter*, the class agrees to provide code for every method in that interface. It is possible for a class to implement many interfaces. In that case, the class must provide all the methods defined by all the interfaces.

An interface is not a class, but you can still make references to interfaces, just like you do with classes. However, you can't use an interface with the *new* keyword. You might write code like this:

```
Reporter rptobj;
.
.
.
rptobj = getNextClassA();
rptobj.report(out);
```

So you can treat *Reporter* just like an object, except that you can't instantiate it and it contains no code. However, you can pretend that any object that implements the interface is one of these mythical *Reporter* objects and use a *Reporter* reference to call any of the interface methods (only *report*, in this case).

Because interfaces have no code, they are similar to *abstract classes*. An abstract class is one that contains at least one method marked *abstract*. These methods, like the methods in an interface, have no code associated with them. This makes the class that contains them abstract. You can extend the class, but you can't instantiate it. This is useful when you want to share code among objects, but there is no clear actual base class. For example, suppose you had classes that represented a serial port, a printer port, and an universal service bus (USB) port. You'd like to share code between them, but what's the common base class? Printer ports are not serial ports, nor are they a kind of USB port.

The answer is to make an abstract class (see Listing 2-14) that represents ports in general. It doesn't make sense to instantiate this class because there is no such thing as a generic port. Unlike an interface, abstract classes can contain resusable code.

```
abstract class GenericPort
{
protected byte [] buffer;
protected int buffp;
protected int bufflen;
protected int portnum;
protected int irq;
public void init();
public int getData(byte [] data);  // returns bytes read
public void sendData(byte [] data, int len);
public GenericPort() { buffer=new byte[256]; buffp=0; bufflen=0;}
public byte getByte()
   {
   if (bufflen==0)
      {
      bufflen=getData(buffer);  // read chunk (assume this never
fails)
      buffp=0;
      }
      return buffer[buffp++];
   }
}
```

Exceptions

Java supports a modern idea known as *exception handling*. Simply put, an exception is a way for your code to signal some event to other parts of your program. Java uses exceptions frequently in its own library and you may also use them as part of your own programs.

Often, but not always, an exception indicates an error has occurred. Suppose you are writing a general-purpose routine that performs a simple calculation based on input parameters. The computation might divide by zero, depending on the input parameters. Of course, you could test for a zero denominator before dividing, but what do you do if you detect this condition? You could print an error message, but that presupposes your program can display a message (remember, I said this routine was general-purpose).

A common solution is to return an error code to the calling function. This is not always good, though. What if the calling program is another general routine? It will have to propagate the error condition somehow.

What if the calling program doesn't check for an error condition? You can solve these problems with exceptions.

When some event occurs—like a division by zero—Java throws an exception. Your code can handle the exception by wrapping the code in a *try* block (see Listing 2-15). In this, case there isn't much advantage to using exceptions. However, suppose the equation inside the *try* block called other methods to do its work. Even if code in these other methods divided by zero, the *catch* block beneath the *try* would be activated (unless, of course, the called functions provided their own *try* block). Consider the example in Listing 2-16.

This is the real value to exceptions. It allows code that is interested in some event to handle that event, no matter what caused it. Code that doesn't care about an event can simply ignore the event.

Dividing by zero is an example of an unchecked exception. Since it could happen at almost any time, Java does not force you to handle the exception. If you remove the *try* and *catch* blocks, the code will still compile, but it will cause an abnormal termination of the program.

Many exceptions, however, are checked exceptions. That means that the Java compiler ensures that you handle the exception wherever it may occur. If your method code calls a function that may throw an exception, you have to either mark your method as throwing the same excep-

Listing 2-15
Coding for
Exceptions

```
public class ex
{
public static void main(String[] args)
  {
  int x=0;
  int y=20;
  int z;
  try
    {
    z=y/x;
    }
  catch (Exception e)
    {
    System.out.println("Divide by zero");
    }
  }
}
```

Listing 2-16
Catching an Exception from a Called Function

```java
public class ex
{
static int docomp(int a, int b)
  {
  return a/b;
  }
public static void main(String[] args)
  {
  int x=0;
  int y=20;
  int z;
  try
    {
    z=docomp(y,x);
    }
  catch (Exception e)
    {
    System.out.println("Divide by zero");
    }
  }
}
```

tion or handle it yourself. You indicate which checked exceptions your function may throw by using a *throws* clause.

You can find an example in Listing 2-17. Here, there is a custom exception (*ScaleError*) that extends *Exception*. When the calculation detects the zero divisor, it throws the custom exception, which can be caught by any of the interested callers. Of course, the *docalc* method could catch the divide by zero exception and simply convert it to the special exception by throwing it in the *catch* clause.

Notice that there are multiple *catch* clauses. The first one is the most specific type of exception. The last one catches any *Exception* object including objects that derive from *Exception*. That's why that clause must come last. If it were first, it would match the *ScaleError* exception and the second *catch* clause would never execute. Try removing the *try* and *catch* block and rebuilding the program. You'll find that the compiler rejects the program because it sees that there is an unchecked exception. Of course, you could mark *main* so that it throws a *ScaleError* exception. Then the exception would terminate the program like an unchecked exception.

```
class ScaleError extends Exception
{
// no methods or fields required
}
public class ex
{
static int docalc(int a, int b) throws ScaleError
  {
  if (b==0) throw new ScaleError();
  return a/b;
  }
public static void main(String[] args)
  {
  int x=0;
  int y=20;
  int z;
  try
    {
    z=docalc(y,x);
    }
  catch (ScaleError e)
    {
    System.out.println("Scale Error");
    }
  catch (Exception e)
    {
    System.out.println("Unknown exception");
    }
  }
}
```

Packages and CLASSPATH

One of Java's biggest features is that it can load class files at runtime. However, that means that, when you create an object, Java has to locate the file that contains the code corresponding to that object.

When Java must locate a class file, it searches the directories listed in the CLASSPATH environment variable. For Windows and similar systems, this is a list of directories separated by semicolons. For Unix-style systems, the directories are separated by colons.

Even with multiple directories, you'd quickly clutter each directory with class files. For that reason, Java supports packages. Packages are somewhat like subdirectories that contain class files. For example, suppose your CLASSPATH variable contains a single directory named C:\Classes. When you attempt to load an ordinary class, the JVM will search in the C:\Classes directory.

However, some classes belong to a package—a group of related classes. For example, you might want to refer to a *HashTable* (a type of associate array provided in the Java library). The *HashTable* class is in the *java.util* package, so to declare it, you'd write

```
java.util.HashTable tbl = new java.util.HashTable();
```

The JVM would look for the HashTable.class file in a subdirectory of one of the CLASSPATH directories. In this case, there is only one directory (C:\Classes) so the class file should be in C:\Classes\java\util\HashTable.class. Of course, if there were more directories listed in the CLASSPATH variable, the JVM would also search those directories, always looking in the java\util subdirectory.

Another way classes can be packaged is in JAR files (that is, a file with a .jar extension). These files are really a type of ZIP (compressed) file that can contain a directory structure and files. You can name JAR files in the CLASSPATH variable, and the JVM will treat them as though they make up a directory. In fact, all of the standard Java library is really in a file named CLASSES.ZIP.

It wouldn't be very convenient to have to write *java.util.HashTable* every time you wanted to use it. By default, if you use a class name, it can only reside in one of the top-level CLASSPATH directories or in the special package *java.lang.* However, you can use the *import* statement to mark certain packages that you want to behave as though they were local.

If you wanted to use the name *HashTable* instead of *java.util.HashTable*, you can add the following line at the start of your Java source file:

```
import java.util.HashTable;
```

You can also get all the classes in java.util by writing:

```
import java.util.*;
```

Keep in mind that you never have to use *import.* If you prefer, you can simply use fully qualified class names everywhere. Still, using *import* makes your programs much more readable so you'll want to use it where appropriate. A common mistake beginning Java programmers make is to try something like this:

```
import System.out.println;
println("Hello World");
```

This won't work! That's because *System* is an object (part of the *java.lang* package), but *out* is a static field of this object. This field is an object reference that has a method called *println*. The *import* statement only works with classes. You can't import a field or method.

Threads

Many programs spend a lot of time waiting for something. If your program doesn't have to do anything else while it's waiting—and it's the only program on the machine—it can just spin in a loop. Many programs, however, are trying to do more than one thing at a time. Even a single-tasking program might be sharing the processor with other programs (especially on a PC or other general-purpose computer).

Threads allow your program to do more than one thing, apparently at the same time. Of course, most computers only have one processor, so really the tasks do not happen simultaneously, but instead share time on the single processor. Each separate thing your program wants to do simultaneously is a thread. Very often, your threads are waiting for something (a network connection, for example) while the main part of your program continues execution.

Each thread is—not surprisingly—an object. The object must either extend the *Thread* object or implement the *Runnable* interface. In either case, you'll provide a *run* method that implements the programming logic you want that thread to execute.

If you derive your object from *Thread,* you can call the object's *start* method to start the thread. If you don't extend *Thread,* you'll have to pass an instance of your object to the constructor of a *Thread* object. Either way, you call *start.* Don't call *run* directly (a common mistake).

When you have multiple threads accessing the same variables, it is crucial that multiple threads don't try to alter variables at the same time. For example, suppose two threads want to increment an integer. The current value of the integer is 10. The first thread might read the value and then lose its time slice (or, on a multiprocessor system, the second thread might be running at the very same time). The second thread might then read the integer, also reading the value 10. Then, when each thread increases the value and writes it back to the variable, they both write the value of 11! Not the intended result.

To prevent this problem, you can declare the field in question as private and only access it through methods. Then you can declare each method as *synchronized.* Each object has a lock associated with it. When any thread attempts to access a *synchronized* member, it must first acquire the associated object's lock. Once one thread owns the lock, no other thread may acquire the lock until the owning thread releases it (which happens automatically when the synchronous operation completes). You can even make static members *synchronized.* Then, the lock affects all objects of a given class, not just one particular instance.

Another related keyword is *volatile.* This keyword marks a variable that may change unexpectedly. The compiler uses this to make sure it doesn't reuse a previous value that might not be current. In normal programming, you don't use this keyword very often. However, in embedded programming *volatile* is very important—hardware registers, for example, might change at any time independent of your program's execution.

Sometimes you don't want to lock an object for the entire duration of a method call. For these cases, you can explicitly synchronize on an object. For example, suppose you want exclusive use of an array named *ary.* Remember that arrays are really objects, and look at this code:

```
synchronized (ary) { ... }
```

Since every object extends *Object,* they all have *wait* and *notify* methods. When you call some object's *wait* method, your thread goes into an efficient waiting state. It will not wake up again until some other thread calls the object's *notify* method. This is very useful for waiting for another thread to perform some task. Of course, both calls must occur in *synchronized* methods (or, at least, a synchronized code block) to ensure threads don't conflict with each other. Calling *wait* automatically releases the lock on the object (otherwise, no other thread could ever call *notify*).

When you call *wait,* you can optionally specify a timeout. If the timeout expires, the *wait* function will return even though no other thread called *notify.*

If more than one thread is waiting on an object, calling *notify* will release one thread (and you can't predict which one it is). However, you can call *notifyAll* to release all threads waiting.

You can assign threads priority (using *setPriority*). Of course, the exact scheduling algorithm depends on the specific JVM you are using (along with the computer and operating system). The *Thread* object provides several static methods you can use to help control execution. If you want it to give you what's left of the current thread's time slice, you can call *Thread.yield*. Of course, if no other thread is ready to execute, the JVM may return control to the current thread immediately anyway.

You can also call *Thread.sleep* to suspend the current thread for a fixed amount of time. Of course, the thread may sleep longer than you requested due to clock resolution limits and scheduling. You can also stop a thread by calling its *suspend* method. The thread will halt until your program (presumably in another thread) calls *resume*.

The *notify* method is useful for allowing one thread to notify other threads that something happened. However, sometimes you want one thread to wait for another thread to complete. You can do this with the *join* method. When you call a *Thread* object's *join* method, your thread will suspend until the thread completes. You can optionally specify a timeout period, if you don't want to wait indefinitely.

Listing 2-18 shows a program that uses two threads (one that extends the *Thread* object, and another that simply implements *Runnable*). As it

Listing 2-18
Program Using Two
Threads

```
class Task2  implements Runnable // could extend from something
{
String msg[]; // message to print
volatile boolean complete=false;
synchronized void init() throws InterruptedException
  {
  while (!complete) wait();
  }
public void run()
  {
  try
    {
    init(); // wait for main to init us
    }
  catch (InterruptedException e) { }
  while (true)
    {
    int i;
    for (i=0;i<msg.length;i++)
      System.out.print(msg[i]);
```

Listing 2-18
Continued

```java
        System.out.print("\n");
      }
    }
  }
public class Thrd extends Thread
  {
  public void run()
    {
    while (true)
      {
      System.out.print("Rise");
      System.out.print("To");
      System.out.print("Vote");
      System.out.println("Sir.");
      }
    }
  synchronized void inittask(Task2 tt2)
    {
    tt2.msg=new String[4];
    tt2.msg[0]="A man";
    tt2.msg[1]="A plan";
    tt2.msg[2]="A Canal";
    tt2.msg[3]="Panama!";
    tt2.complete=true;
    tt2.notify();
    }
  public static void main(String [] args)
    {
    Thrd t1=new Thrd();
    Task2 tt2=new Task2();
    Thread t2=new Thread(tt2);
    t1.start();
    t2.start();
    t1.inittask(tt2);
    try
      {
      t1.join();  // wait forever
      }
    catch (Exception e)
      {
      }
    }
  }
```

exists, the first thread initializes the second thread after it is running. That means the second thread must wait for the first thread to notify it. Even then, the output from each thread is interspersed with output from the other thread.

To correct this problem, you'd need to synchronize the two output routines. You'll find one possible way to do this in Listing 2-19. Here, a simple object (*OutputLock*) synchronizes the two threads' access to the output stream. The object has a static member that returns a single instance of the *OutputLock* function. The threads use *synchronize* to lock out the other thread (or threads). Now the output appears as you expect.

Listing 2-19
Synchronizing Two
Output Routines

```
class OutputLock // class just to act as lock
  {
  private static OutputLock instance;
  static OutputLock getInstance()
    {
    if (instance==null) instance=new OutputLock();
    return instance;
    }
  }
class Task2   implements Runnable // could extend from something
  {
  String msg[]; // message to print
  volatile boolean complete=false;
  synchronized void init() throws InterruptedException
    {
    while (!complete) wait();
    }
  public void run()
    {
    try
      {
      init(); // wait for main to init us
      }
    catch (InterruptedException e) { }
    while (true)
      {
      int i;
```

Listing 2-19
Continued

```java
        synchronized (OutputLock.getInstance())
          {
            for (i=0;i<msg.length;i++)
              System.out.print(msg[i]);
            System.out.print("\n");
          }
        }
      }
    }
public class Thrd extends Thread
  {
  public void run()
    {
    while (true)
      {
      synchronized (OutputLock.getInstance())
        {
        System.out.print("Rise");
        System.out.print("To");
        System.out.print("Vote");
        System.out.println("Sir.");
        }
      }
    }
  synchronized void inittask(Task2 tt2)
    {
    tt2.msg=new String[4];
    tt2.msg[0]="A man";
    tt2.msg[1]="A plan";
    tt2.msg[2]="A Canal";
    tt2.msg[3]="Panama!";
    tt2.complete=true;
    tt2.notify();
    }
  public static void main(String [] args)
    {
    Thrd t1=new Thrd();
    Task2 tt2=new Task2();
```

Listing 2-19
Continued

```
    Thread t2=new Thread(tt2);
    t1.start();
    t2.start();
    t1.inittask(tt2);
    try
      {
      t1.join();  // wait forever
      }
    catch (Exception e)
      {
      }
      }
    }
```

SUMMARY

You could read an entire book on Java—there are plenty around. However, this chapter, along with the examples in the next few chapters, will give you a lot of practice with Java. You can also find many online tutorials, books, and documentation on Java. Be sure to check out the online resources section for more information. Be aware, though, that many books and other materials will focus on writing graphical programs, not embedded systems.

This chapter may leave you wondering why you should use Java. In the next chapter, however, you'll see that Java's networking capability is a real winner. Also, Java's cross-platform ability will serve you well in a networked environment.

ONLINE RESOURCES

http://java.sun.com. Java's home on the Web. Free downloads of the JDK, tutorials, news, and more.

http://www.norvig.com/java-iaq.html. Java Infrequently Asked Question (IAQ) list.

http://mindprod.com/gotchas.html. Java gotchas.

http://www.afu.com/javafaq.html. Java programmer's FAQ.

A PC Gateway

A few years ago, during a cross-country trip, my van blew a head gasket somewhere in the middle of Arkansas. Not too many years ago, this would have necessitated walking miles to try to find help. Of course, today you just take out your trusty cell phone (which works even in the middle of Arkansas) and summon help. Of course, the problem I found was that I didn't know exactly where I was! I had to walk at least far enough to find a mile marker so the tow truck could find us. The resulting week-long stay at a hotel in a very small town left us tired of eating at the lone truck stop and glad to get back on the road with our wallet much lighter than it had been before.

Embedded Internet systems have a similar problem. You not only need a phone (a connection to the Internet), but you also need to know where you are (an Internet address). There are a variety of ways you might connect a system to the Internet:

1. The most robust way to connect to the Internet is through an Ethernet port. At first, this sounds like overkill for an embedded system, but actually Ethernet transceivers on a chip are common and inexpensive. The biggest problem is actually finding a place to plug the Ethernet connection into the Internet. Perhaps your device will be in a school or office building that has a network. Many cable and DSL modems use Ethernet ports, as well.

2. The traditional method of connecting to the Internet for small devices is via a modem. Again, you might not think about using a consumer-style modem in an embedded environment. However, there are several modem modules that make it very easy to add a modem to a device. In addition, modem chip sets are inexpensive and plentiful, although modem modules are more cost-effective unless you are building many identical devices.

3. Another way to connect to the Internet is to use a private connection to another computer that has an Internet connection (a gateway). For example, your controller might use a serial port to talk to a PC that connects to the Internet in the usual way.

Throughout this book, you'll see examples of each of these techniques. However, this chapter will show you how to use a PC as a gateway. This will let you practice Java on a workstation before trying it on an embedded system.

At first glance, you might worry that connecting via a PC is inefficient. However, if you are trying to adapt existing devices to the net, it might not be as inefficient as you think. First, a modern PC can service several devices with ease. Second, small embedded PCs are now available and relatively inexpensive. Beside this, PC development tools are powerful and widely available. Many older devices sent their data to a PC anyway—all you need to do is funnel that data to the Internet. Of course, you won't have a PC connecting your toaster to the Internet, but for many industrial applications, this is a very useful way to make a connection.

By itself, Java doesn't have facilities to read the serial port. However, Sun provides an extension called javax.comm that can easily handle communications with serial and parallel ports. You'll have to download this package separately from the JDK, and it is not available for all platforms.

Object Power

As you read about the javax.comm package, you'll see how Java's object orientation works to your advantage. Dealing with the serial port is a difficult task, and doing so varies greatly between different types of computers. However, if your computer supports the javax.comm package, you'll be able to easily access ports regardless of the platform. That's because javax.comm provides objects that conceal the details about the ports and the platform idiosyncrasies.

In this chapter, you'll see how to use a PC-based Java program to read data from a simple microcontroller circuit that is not Internet-aware. The PC will then e-mail you the results from the microcontroller.

Another use of objects in this program is the e-mail sending portion of the code. To see how the program works, you can just assume that the Simple Mail Transfer Protocol object (*SMTP* object) provided will send e-mail correctly. You don't need to know all the details about its operation. But after we've used the *SMTP* object, I'll show you how it does work, because it is an excellent way to learn how to write networking code in Java.

Using Serial Input/Output

If you have the javax.comm package installed correctly, using the serial port is easy. Your program simply imports javax.comm.* and then it can access several classes you'll need to work with the port.

The first class you'll use is *CommPortIdentifier*. This class is necessary because each platform has different names for the available ports. There are several ways you can use this class, but of particular use is the static method *getPortIdentifiers*. This returns an *Enumeration* object (more precisely, an object that implements the *Enumeration* interface). This enumeration will return *CommPortIdentifier* objects for each known port on the computer. To list the ports on your computer, you could use the following code:

```
Enumeration e;
e=CommPortIdentifier.getPortIdentifiers();
while (e.hasMoreElements())
    {
      CommPortIdentifiercid=(CommPortIdentifier)e.nextElement();
      System.out.println(cid.getName());
    }
```

Of course, not all ports are serial ports. If you wanted to restrict the list to serial ports only, you could change the *println* function to

```
if (cid.getPortType()==CommPortIdentifier.PORT_SERIAL)
    System.out.println(cid.getName());
```

If you know the port's name, you can use it directly by calling *CommPortIdentifier.getPortIdentifier*. Be careful though; the port names on your development system (like COM1) may not match up with the target system. For example, on a Unix or Linux system, serial devices will probably appear as a file in the /dev directory.

Opening a Port

Once you have the *CommPortIdentifier* you want, you can use it to open the port. You simply call the object's *open* method. This returns a *CommPort* object. If you are working with a port you know is a serial port, you'll want to cast this to a *SerialPort* object (which, of course, derives from *CommPort*). Here's an example:

```
SerialPort port=(SerialPort)cid.open("SerialDemo",2000);
```

The first argument to *open* is an identifying name—usually the name of your program. The second argument is a timeout value. If the port is in use, Java will ask the program using the port to relinquish it. If the program doesn't release the port within the timeout period (2000 ms in this case), the *open* call will fail.

By default, the port will use 9600 baud, 8 bits, 1 stop bit. However, you can call *setSerialPortParams*. This call can throw an exception if your hardware doesn't support the parameters you've requested.

Once you have a port, you'll want to read and write data. There are several ways to do this, but all of them will involve Java streams.

Stream Basics

Java's input/output (I/O) system can be somewhat daunting at first. The idea is that you start with a basic stream (for example, an *InputStream*). In this case, the *InputStream* is from the *SerialPort* object's *getInputStream* method. However, there are many other places you can get a stream, including the console input stream in *System.in* and the corresponding output stream in *System.out*.

If all you need is to read and write bytes, you can just use the stream directly. However, you will often want to read or write in more sophisticated ways. For example, you might want to read Java objects or lines instead of bytes (remember, Java strings are not composed of bytes). Java provides many I/O stream classes to handle these special cases. You'll construct these objects and pass the base stream as an argument to the constructor. Sometimes, you'll use the new object as an argument to yet another specialized stream class.

For example, suppose you want to read lines of text. The class that can do this is *BufferedReader*, which contains the *readLine* method. The constructor for *BufferedReader* requires a *Reader* object. Unfortunately, *getInputStream* doesn't return a *Reader*, it returns an *InputStream*. You'll have to convert the *InputStream* to an *InputStreamReader* (which derives from *Reader*). Then you can convert the *InputStreamReader* into a *BufferedReader*. Your code might look like this:

```
InputStream portis = port.getInputStream();
InputStreamReader isr = new InputStreamReader(portis);
Buffered Reader portrdr = new BufferedReader(isr);
```

You could also dispense with the intermediate variables and be a bit more succinct:

```
BufferedReader portrdr = new BufferedReader(new
InputStreamReader(port.getInputStream()));
```

There are other classes you might use for other specialized cases. For example, *DataInputStream* allows you to read basic types like integers or floats in a machine-independent way. If you wanted to keep track of how many lines you've read, you might elect to use *LineNumberReader* instead of *BufferedReader*. This class extends *BufferedReader*, but it also keeps track of how many lines it has read and makes that count available via the *getLineNumber* method.

The beauty of this method is that you can apply the additional processing to any sort of stream. For example, in this case, the program's input is from a port. However, you could easily make a *FileReader* to read from a file. Other input streams might originate from an array of characters, the console, or a network socket. Regardless, you can always attach the stream to classes like *BufferedReader*.

Output works the same way. You might, for example, attach an output stream to a *PrintWriter* to convert Java variables into human-readable form on output (similar to *System.out*).

Advanced Serial I/O

You might not want your program to wait for input to arrive—you may want it to do other processing instead of just waiting. Of course, one way to accomplish this would be to wait for input in a thread. However, you can also use the port's *addEventListener* to add an object that implements the *SerialPortEventListener* interface. You can add this interface to your main class by writing a single method (*serialEvent*). Of course, you also have to use the *implements* keyword to inform Java that you are providing this interface.

Java can call *serialEvent* when many things happen. For example, you might want the call to occur when data is available, or when the carrier detect or other handshaking line changes state. You register for these events by calling methods like *notifyOnDataAvailable*. That way Java doesn't bother calling *serialEvent* for events you don't plan to handle.

Speaking of events, you can also register with the *CommPortIdentifier* object to notify you when the port's ownership status changes. You'll need to implement the *CommPortOwnershipListener* interface (a single function). Java will call your function when someone opens or closes the port. In addition, if another program tries to open the port while your program has it open, this interface receives a *PORT_OWNER-SHIP_REQUESTED* code. If you close the port, it will allow the other program to proceed.

An Example

Consider the schematic in Fig. 3-1. This circuit uses a Basic Stamp (a microcontroller that programs in Basic) to read a thermistor (or, for that matter, any resistive sensor). The simple basic program (see Listing 3-1) reads the sensor regularly and writes the result to the Basic Stamp's serial port.

Suppose this simple device is an existing device that you want to place on the Internet. A serial port gateway can accomplish this with a minimum of effort. The data format is an asterisk, followed by four hex digits, and then a carriage return and line feed.

The program in Listing 3-2 will run on a PC or other platform with the javax.comm extensions. It reads the data from the Basic Stamp circuit and examines the value it receives. If the data appears to be in the correct format, the program strips the hex value out of the received packet. It then compares the value to the last value. If the difference is greater than 10, the program sends an e-mail to a predetermined e-mail address.

Figure 3-1
Circuit Using Basic Stamp

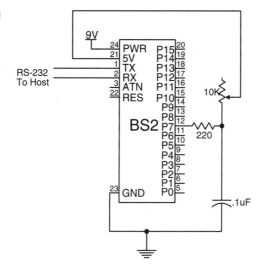

Listing 3-1
Simple Basic Stamp
Program

```
pot con 7      ' sensor input
sport con 16  ' serial port
baud con 84   ' baud rate constant (84 is 9600 8 N 1)
counts var word
delay con 60000  ' 1 minute
top:
 high pot  ' discharge pot
 pause 1
 rctime pot,1,counts
 serout sport,baud,["*",hex4 counts,13,10]
' debug "*",hex4 counts,cr
 pause delay
 goto top
```

```
import javax.comm.*;
import java.util.*;
import java.io.*;
// VERY IMPORTANT: On Windows some versions of javax.comm
// Require that your program and the win32com.dll file
// be on the same drive! This is a bug, but if
// your computer seems to have no ports, this
// is probably the problem.
public class SerialDemo
{
  public static void main(String args[]) throws Exception
  {
      Enumeration e;
      String portname;
      Date today;
      int rc;
      CommPortIdentifier cid;
      SMTP smtp=new SMTP("bardo.clearlight.com");
      if (args.length==0)
```

Listing 3-2
Program with javax.comm Extension

```
        {
        e=CommPortIdentifier.getPortIdentifiers();
        System.out.println("Your system has the following serial ports:");
        while (e.hasMoreElements())
            {
                cid=(CommPortIdentifier)e.nextElement();
                if (cid.getPortType()!=CommPortIdentifier.PORT_SERIAL)
                    continue;
                System.out.println(cid.getName());
            }
        System.out.println("Please select one:");
        // read port
        BufferedReader rdr = new BufferedReader(
            new InputStreamReader(System.in));
        portname=rdr.readLine();
        }
    else
      {
        portname=args[0];
      }
    cid=CommPortIdentifier.getPortIdentifier(portname);
// commented out is another way to search all the ports...
    //      e=CommPortIdentifier.getPortIdentifiers();
    //        while (e.hasMoreElements())
    //        {
    //              CommPortIdentifier cid
    //                = (CommPortIdentifier)e.nextElement();
    //              if (cid.getPortType()
    //                   !=CommPortIdentifier.PORT_SERIAL) continue;
    //              if (portname.equals(cid.getName()))
    //              {
    // open port
            SerialPort port=(SerialPort)cid.open("SerialDemo",2000);
            BufferedReader portrdr = new
             BufferedReader(new InputStreamReader(
              port.getInputStream()));
```

Listing 3-2
Continued

```
            MailMessage msg;
            int lastvalue=0;
            boolean first=true;
// wait for data from Stamp
            System.out.println("Listening...");
            while (true)
                {
                    int value;
                    String inline = portrdr.readLine();
                    // expect *XXXX\n where XXXX is hex
                    System.out.println(inline);
                    if (inline.charAt(0)!='*') continue;
                    value=Integer.parseInt(inline.substring(1),16);
                    // don't e-mail if within 10
                    if (!first)
                            if (Math.abs(value-lastvalue)<10) continue;
                    System.out.println("Sending " +
                        Math.abs(value-lastvalue));
                    first=false;
                    lastvalue=value;
// Send e-mail message
            today=new Date();
            msg=new MailMessage("alw@al-williams.com",
                        "instrument@al-williams.com",
                        "Instrument reading",
                        today.toString()+": "+value);
            rc=smtp.sendMail(msg);
            if (rc!=0)
                {
                    System.out.println("Error: " +
            smtp.getLastResponse());
                }
            }
        //   }
    //   }
    //   System.out.println("Sorry. I can't find that port");
    }
}
```

Listing 3-2
Continued

Here are a few important things to notice about Listing 3-2:

- The entire logic of the program is in the *main* function. The only objects in use are those provided by Java and the *SMTP* object I will show you shortly.

- The *main* function examines the command line arguments for a port name. If it finds one, it attempts to open it directly. If it does not, the program prints a list of available ports and allows the user to select one.

- To display the ports, the program creates an *Enumeration* object. In reality, *Enumeration* is an interface that objects that provide lists implement. The program doesn't know or care what type of object is really in use (it is some internal object provided by javax.comm). The only requirement is that the object implements *Enumeration*. Walking an *Enumeration* with *hasMoreElements* and *nextElement* is a common operation in Java programming. Each call to *nextElement*, of course, sets the object to return the next element on the next call.

- When selecting a port manually, the program must ask for input using *System.in* (the console input stream). Notice that the program converts the stream to a *BufferedReader* so that it can read an entire line at once.

- The program reads the number and compares it to the previous reading. The *Math.abs* function is a static member of the *Math* object and computes the absolute value.

- Each e-mail message has a time stamp recorded. The *Date* object, if constructed with no arguments, creates an object that reflects the date and time at creation.

- The *SMTP* object uses a *MailMessage* object to represent an e-mail message. Calling *SendMail* transmits the message.

Tip: If you are trying this program on the PC, be wary. At least some versions of the javax.comm package from Sun have a subtle bug. If you are running your program from the same disk drive as the Java libraries, you won't have any problems. However, if the program is on a different disk drive, the library will report that no serial ports are available on the PC. For example, if your Java libraries are on your D: drive and your program is on your E: drive, you'll see this failure.

■■ ■■ SMTP

Ordinarily, sending e-mail would be a chore and would require some research to successfully get it working. However, provided with a ready-made class, it becomes trivial. You simply construct the *SMTP* object (providing the name of an SMTP server), build a *MailMessage* object, and call *SendMail*. The object takes care of all the details. However, looking inside the *SMTP* class is a good way to get familiar with Java networking—something you need to know anyway.

Inside SMTP

When you send an e-mail, you connect to an SMTP host computer on port 25. This host delivers mail to local users, and forwards it on if the recipient is a user on another computer. You can find all the technical details about SMTP in the Internet RFC821 (Online Resources at the end of the chapter) dating back to 1982. Reading the protocol, you get the idea that the author wanted to allow someone to dial into an SMTP host using a TeleType (this is 1982, remember) and manually conduct business with the software.

However, the author also allows for one SMTP host to forward mail to another. So the protocol is amenable to machine interpretation as well as manual entry. If you use a telnet program to log into an SMTP host (on port 25), you'll see a response like this one:

```
220 smtp1b.mail.yahoo.com ESMTP
```

This response is from the Yahoo SMTP server at smtp.mail.yahoo.com.

Every line the host sends begins with a three-digit response code. The fourth character is a hyphen if there are more lines related to this response forthcoming. If the fourth character of the line is a space, then this line is the last line of a multiline response, or—as is usually the case—the only line. Programs usually only read the first four characters and ignore the rest of the line. As you might expect, the rest of the line is more useful than the codes for humans. You can find a list of response codes and their meaning in Table 3-1.

There are many possible commands you can send an SMTP server (see Table 3-2). However, for the purposes of this program, all you need are a few simple commands (the ones in bold in Table 3-2). The order of the commands is important. The basic sequence will be

<table>
<tr><td>**TABLE 3-1**</td><td colspan="2">Code   Description</td></tr>
</table>

	Code	Description
TABLE 3-1 SMTP Response Codes Rely on Three-Digit Numbers to Identify Themselves; the Server Will Also Provide a Text Message That Most Automated Programs Ignore	211	System status, or system help reply
	214	Help message
	220	\<domain\> Service ready
	221	\<domain\> Service closing transmission channel
	250	Requested mail action okay, completed
	251	User not local; will forward to \<forward-path\>
	354	Start mail input; end with \<CRLF\>.\<CRLF\>
	421	\<domain\> Service not available, closing transmission channel
	450	Requested mail action not taken: mailbox unavailable
	451	Requested action aborted: local error in processing
	452	Requested action not taken: insufficient system storage
	500	Syntax error, command unrecognized
	501	Syntax error in parameters or arguments
	502	Command not implemented
	503	Bad sequence of commands
	504	Command parameter not implemented
	550	Requested action not taken: mailbox unavailable
	551	User not local; please try \<forward-path\>
	552	Requested mail action aborted: exceeded storage allocation
	553	Requested action not taken: mailbox name not allowed
	554	Transaction failed

HELO—Identification of the client

MAIL FROM:—Identification of the sender

RCPT TO:—Identification of recipient (may occur more than once)

DATA—Specification of e-mail message

QUIT—Termination of session with server

Each command has only one or two possible responses that are not errors. Most e-mail programs allow you to specify multiple recipients, as

Command	Description
DATA	Begins the actual contents of the e-mail message (ends with a line containing only a period)
EXPN	Expands a mailing list
HELO	Identifies sender via host name or IP address
HELP	Sends a help response
MAIL FROM:	Starts a mail transaction from a sender
NOOP	Do nothing
QUIT	End session
RCPT TO:	Indicates a single recipient for the e-mail
RSET	Aborts current transaction
SEND	Sends data directly to a terminal
SOML	Sends data to a terminal or via e-mail
SAML	Sends data to a terminal and via e-mail
TURN	Causes the sender to become the receiver and vice versa
VRFY	Verifies an e-mail address

well as carbon copy recipients and blind carbon copy recipients (CC and BCC). As far as SMTP cares, all of these are recipients that you specify with separate *RCPT TO:* commands.

The difference between a CC recipient and a BCC recipient is if the recipient appears in the e-mail's headers or not. Like a Web document, e-mail has a header and a body (you can find the details in RFC822; see Online Resources at the end of this chapter). You can send many headers, including the subject, a reply path, and other headers that particular e-mail programs might interpret in different ways. SMTP doesn't care at all which headers you include (if any).

Timeouts, Multiple Lines, and Transparency

When I looked at some other SMTP examples on the Web, I noticed that many of them did not follow the RFC completely. They would work in most cases, but a few odd conditions might cause failure. I tried to avoid these in my code. In particular, I noticed that many examples didn't correctly handle the following situations:

- *Timeout.* It is possible for any number of reasons that the server might fail to respond. It is prudent to allow the server a certain amount of time to respond and then terminate the connection. Some codes didn't allow for this possibility. Others did allow for timeouts, but did so inefficiently by simply looping.

- *Multiple lines.* As I mentioned earlier, some servers will reply with multiple lines for a single response. Each response begins with the three-digit code, but the fourth character will be a hyphen for all but the last line. Some examples of SMTP code do not consider this.

- *Transparency.* When sending mail messages, the SMTP server looks for a line containing a single period to terminate the message. However, what if the e-mail message itself contains a single period? To prevent this problem, SMTP clients should add a period to all lines that start with a period. If the server sees a line that begins with two periods, it strips the first one off automatically.

Implementation

Before tackling the SMTP class itself, I wrote two small helper classes (see Listings 3-3 and 3-4). The first, *MailMessage*, simply encapsulates the strings that make up a typical e-mail message. The second class (*SMTPResults*) is simply a container for static integers that represent the response codes from Table 3-1.

```
/**
 * This is just a collection of data that makes up a mail message.
 * @author Al Williams
 * @version 1.0
 * @see SMTP#sendMail
 */
public class MailMessage
```

Listing 3-3
Small Helper Class *MailMessage*

```
{
  /**
   * Default constructor.
   * @param None
   */
  public MailMessage()
  {
  }
  /**
   * Constructor that initializes.
   * @param _from The sender's e-mail address.
   * @param _to The recipient's e-mail address (separate multiples with ;)
   * @param _cc Carbon copy addresses (; separate multiples).
   * @param _bcc Blind copy addresses (; separate multiples).
   * @param _subject The e-mail subject.
   * @param _body The body of the e-mail.
   */
  public MailMessage( String _from, String _to,
  String _cc, String _bcc, String _subject, String _body )
  {
      sender = _from;
      to = _to;
      cc = _cc;
      bcc = _bcc;
      subject = _subject;
      body = _body;
  }
  /**
   * Constructor with commonly required arguments.
   * @param _from The sender's e-mail address.
   * @param _to The recipient's e-mail address (use ; between addresses).
   * @param _subject The e-mail subject.
   * @Param _body The body of the e-mail.
   */
  public MailMessage( String _from, String _to, String _subject, String _body )
```

Listing 3-3

Continued

```
{
    sender = _from;
    to = _to;
    subject = _subject;
    body = _body;
}
/**
 * The sender's e-mail address.
 */
public String sender;
/**
 * The recipient's e-mail address (separate multiple addresses with semicolons).
 */
public String to;
/**
 * The Carbon Copy addresses (separate multiple addresses with semicolons).
 */
public String cc;
/**
 * The Blind Carbon Copy addresses
 * (separate multiple addresses with semicolons).
 */
public String bcc;
/**
 * The e-mail's subject.
 */
public String subject;
/**
 * The e-mail's body.
 */
public String body;
}
```

Listing 3-3
Continued

```java
/**
 * Static values for SMTP result codes
 * @author Al Williams
 * @version 1.0
 */
public class SMTPResults
{
    /**
     * Syntax error, command unrecognized
     */
    static final public int SMTP_RESULT_UNRECOG = 500;
    /**
     * Syntax error in parameters or arguments
     */
    static final public int SMTP_RESULT_PARAM = 501;
    /**
     * Command not implemented
     */
    static final public int SMTP_RESULT_UNIMPLEMENTED = 502;
    /**
     * Bad sequence of commands
     */
    static final public int SMTP_RESULT_SEQUENCE = 503;
    /**
     * Command parameter not implemented
     */
    static final public int SMTP_RESULT_PARAMNI = 504;
    /**
     * System status, or system help reply
     */
    static final public int SMTP_RESULT_SYSTEM = 211;
    /**
     * Help message
     */
    static final public int SMTP_RESULT_HELP = 214;
```

Listing 3-4
Small Helper Class *SMTPResults*

```
/**
 * <domain> Service ready
 */
static final public int SMTP_RESULT_READY = 220;
/**
 * <domain> Service closing transmission channel
 */
static final public int SMTP_RESULT_CLOSING = 221;
/** <
 * domain> Service not available, closing transmission channel
 */
static final public int SMTP_RESULT_SERUNAVAILABLE = 421;
/**
 * Requested mail action okay, completed
 */
static final public int SMTP_RESULT_COMPLETED = 250;
/**
 * User not local; will forward to <forward-path>
 */
static final public int SMTP_RESULT_FORWARD = 251;
/**
 * Requested mail action not taken: mailbox unavailable
 */
static final public int SMTP_RESULT_MBXUNAVAILABLE = 450;
/**
 * Requested action not taken: mailbox unavailable
 */
static final public int SMTP_RESULT_NOTTAKEN = 550;
/**
 * Requested action aborted: error in processing
 */
static final public int SMTP_RESULT_ABORTED = 451;
/**
 * User not local; please try <forward-path>
 */
```

Listing 3-4

Continued

```
static final public int SMTP_RESULT_USER_NOT_LOCAL = 551;
/**
 * Requested action not taken: insufficient system storage
 */
static final public int SMTP_RESULT_STORAGE = 452;
/**
 * Requested mail action aborted: exceeded storage allocation
 */
static final public int SMTP_RESULT_EXSTORAGE = 552;
/**
 * Requested action not taken: mailbox name not allowed
 */
static final public int SMTP_RESULT_NOT_ALLOWED = 553;
/**
 * Start mail input; end with <CRLF>.<CRLF>
 */
static final public int SMTP_RESULT_MAIL_START = 354;
/**
 * Transaction failed
 */
static final public int SMTP_RESULT_TRANS_FAILED = 554;
}
```

Listing 3-4
Continued

The main class, *SMTP* (see Listing 3-5), has only two important methods. The constructor you'll use to create an instance of *SMTP* requires a host name or IP address of an SMTP server. The other method is *sendMail*. You simply pass a filled-in *MailMessage* object to *sendMail* and it does the rest. If *sendMail* returns 0, then everything went well. If an error occurs, you'll find the SMTP response code in the return value.

The internal details of *SMTP* are a bit more interesting. The *sendMail* function simply opens a socket and then calls *sendMailEngine*. That way, even if *sendMailEngine* returns an error, the *sendMail* function can properly close the socket. The *sendAddress* method handles separating e-mail addresses to send separate RCPT TO: commands.

```java
import java.awt.*;
import java.net.*;
import java.util.*;
import java.io.*;
/**
* SMTP Class. You can use this class to
* send e-mail via an SMTP server.
* You can find more info about e-mail and SMTP via the RFCs particularly
* <A HREF=http://www.faqs.org/rfcs/rfc821.html>RFC821</A> and
* <A HREF=http://www.faqs.org/rfcs/rfc822.html>RFC822</A>
* @author Al Williams
* @version 1.0
*/
public class SMTP
{
    // things you might want to change
    // 30 seconds timeout
    final static int WAIT_TIMEOUT = ( 30 * 1000 );
    final static int smtpPort = 25;
    final static String addressSep = ";";  // separates e-mail addresses
// SMTP server for testing
    final static String testServer = "bardo.clearlight.com";
    String smtpServer;
    // could hardcode this
    MailMessage message;
    String hostname;
    BufferedReader input;
    OutputStream output;
    String errorText;    // copy of last response -- in case of error
    Socket sock;
    final String crlf = "\r\n";
    /**
     * Get the last response message. Useful for displaying SMTP errors.
     * @returns String containing last response from server.
     */
```

Listing 3-5
Main Class, *SMTP*

```
String getLastResponse()
 {
 if (errorText==null) return "Unable to connect or unknown error";
 return errorText;
 }
/**
 * This is a test main — if you run SMTP it will send an e-mail
 * @param args The command line arguments
 */
public static void main( String args [] )
{
    int rc;
    Date today = new Date();
    if (args.length!=1)
    {
      System.out.println("Usage: SMTP e-mail address");
      System.exit(1);
    }
    System.out.println("Sending test message to " + args[0]);
    MailMessage msg = new MailMessage( "alw@al-williams.com",
      args[0], "Test",
     "Sent at " + today.toString());
    SMTP smtp = new SMTP( testServer);
    rc = smtp.sendMail( msg );
    if( rc != 0 )
    {
      System.out.println( "Error " + rc );
      System.out.println(smtp.getLastResponse());
    }
    else
      System.out.println( "OK" );
}
/**
 * Constructor. Requires SMTP server name.
 * @param host An SMTP server. Must be accessible from this code.
```

Listing 3-5
Continued

```
 *           In particular, applets can usually only connect back to the same host
 *           they originated from.
 */
public SMTP( String host )
{
  smtpServer = host;
}
//read data from input stream to buffer String
private int getResponse( int expect1 ) throws IOException
{
   return getResponse( expect1,-1 );
}
synchronized private int getResponse( int expect1, int expect2 )
   throws IOException
{
  boolean defStatus;
  long startTime;
  int replyCode;
  TimeoutRead thread = new TimeoutRead( input );
  thread.setBuffer("");
  thread.start();
  try
  {
     thread.join( WAIT_TIMEOUT );
  }
  catch( InterruptedException e )
  {
  }
  if( thread.isComplete() && thread.getBuffer().length() > 0 )
  {
     try
     {
        errorText = thread.getBuffer();  // if there is an error, this is it.
        replyCode = Integer.valueOf( errorText.substring( 0, 3 ) ).intValue();
        if( replyCode == 0 ) return-1;
```

Listing 3-5
Continued

```
                if( replyCode == expect1 || replyCode == expect2 ) return 0;
                return replyCode;
            }
            catch( NumberFormatException e )
            {
               return-1;
            }
         }
      return-1;
      // nothing in buffer
   }
   private void writeString( String s ) throws IOException
   {
       output.write( s.getBytes() );
   }
   private int sendAddresses(String pfx, String addr) throws IOException
   {
      int n0=0;
      int n;
      int rv=0;
      while (rv==0 && (n=addr.indexOf(addressSep,n0))!=-1)
        {
        writeString(pfx+addr.substring(n0,n)+crlf);
          n0=n+1;
        rv=getResponse( SMTPResults.SMTP_RESULT_COMPLETED,
          SMTPResults.SMTP_RESULT_FORWARD );
        }
      if (rv==0)
      {
        writeString(pfx+addr.substring(n0)+crlf);
        rv=getResponse( SMTPResults.SMTP_RESULT_COMPLETED,
          SMTPResults.SMTP_RESULT_FORWARD );
      }
      return rv;
   }
```

Listing 3-5
Continued

```
/**
 * Use sendMail to actually send an e-mail message.
 * @param MailMessage This is a filled-in MailMessage object that specifies
 *   the text, subject, and recipients.
 * @return Zero if successful. Otherwise, it returns the SMTP return code.
 * @see SMTPResults
 */
public int sendMail( MailMessage msg )
{
   String inBuffer;
   String outBuffer;
   int rv;
   message = msg;
   if( msg.to == null || msg.to.length() == 0 )
   {
      return-1;
   }
   // Create connection
    try
    {
       sock = new Socket( smtpServer, smtpPort );
       hostname = "[" + sock.getLocalAddress().getHostAddress() + "]";
    }
    catch( IOException e )
    {
        return-1;
    }
    //Create I/O streams
    try
    {
       input = new BufferedReader(
         new InputStreamReader( sock.getInputStream() ) );
    }
    catch( IOException e )
    {
```

Listing 3-5
Continued

```
            return—1;
        }
        try
        {
            output = sock.getOutputStream();
        }
        catch( IOException e )
        {
            return—1;
        }
        rv=sendMailEngine();
// end connection
        try
        {
            sock.close();
             sock=null;
        }
        catch( IOException e )
        {
            return—1;
        }
        return rv;
    }
// this is a separate routine so the main sendMail can  always close the socket
private int sendMailEngine()
{
    try
    {
        int replyCode;
        int n;
        Date today = new Date();
        replyCode = getResponse( SMTPResults.SMTP_RESULT_READY );
        if( replyCode != 0 ) return replyCode;
        //Send HELO
        writeString( "HELO " + hostname + crlf );
```

Listing 3-5
Continued

```
replyCode =
  getResponse( SMTPResults.SMTP_RESULT_COMPLETED );
if( replyCode != 0 ) return replyCode;
// Identify sender
writeString( "MAIL FROM: " + message.sender + crlf );
replyCode =
  getResponse( SMTPResults.SMTP_RESULT_COMPLETED );
if( replyCode != 0 ) return replyCode;
// Send to all recipients
replyCode = sendAddresses("RCPT TO: ",message.to);
if( replyCode != 0 ) return replyCode;
// Send to all CC's (if any)
if( message.cc != null && message.cc.length() != 0 )
{
   replyCode = sendAddresses("RCPT TO: ",message.cc);
   if( replyCode != 0 ) return replyCode;
}
   // Send to all BCCs (if any)
   if( message.bcc != null && message.bcc.length() != 0 )
   {
         replyCode = sendAddresses("RCPT TO: ",message.bcc);
      if( replyCode != 0 ) return replyCode;
   }
   // Send message
   writeString( "DATA" + crlf );
   replyCode = getResponse(
      SMTPResults.SMTP_RESULT_MAIL_START );
   if( replyCode != 0 ) return replyCode;
   //Send mail content CRLF.CRLF
   // Start with headers
   writeString( "Subject: " + message.subject + crlf);
   writeString( "From: " + message.sender + crlf);
   writeString( "To: " + message.to + crlf);
   if( message.cc != null && message.cc.length() != 0 )
                  writeString( "Cc: " + message.cc + crlf);
```

Listing 3-5
Continued

```
            writeString( "X-Mailer: SMTP Java Class by Al Williams" + crlf );
            writeString( "Comment: Unauthenticated sender" + crlf );
            writeString( "Date: " + today.toString() +crlf + crlf );
// fix periods for transparency
            StringBuffer body=new StringBuffer(message.body);
            n=0;
            do
            {
             n=body.toString().indexOf("\n.",n);
             if (n!=-1) body.insert(++n,'.');
            } while (n!=-1);
            writeString( body.toString() );   // send body
            writeString( crlf + "." + crlf );  // end mail
            replyCode = getResponse(
              SMTPResults.SMTP_RESULT_COMPLETED );
            if( replyCode != 0 ) return replyCode;
            // Quit
            writeString( "QUIT" + crlf );
            replyCode =
              getResponse( SMTPResults.SMTP_RESULT_CLOSING );
            if( replyCode != 0 ) return replyCode;
        }
        catch( IOException e )
        {
          return—1;
        }
     return 0;
      }
   protected void finalize() throws Throwable
   {
   if (sock!=null) sock.close();
   super.finalize();
   }
   }
   // private class to handle the reading in a thread
```

Listing 3-5
Continued

```
class TimeoutRead extends Thread
{
   private String buffer = new String( "" );
   private BufferedReader input;
   private boolean complete = false;
   synchronized String getBuffer() { return buffer; }
   synchronized void setBuffer(String s) { buffer=s; }
   synchronized public boolean isComplete()
   {
     return complete;
   }
   public TimeoutRead( BufferedReader i )
   {
      input = i;
   }
   public void run() // do input in thread
   {
     try
     {
        do
         {
            setBuffer(input.readLine());
         } while (getBuffer().charAt(3)=='-'); // loop on multiline response
// This line is useful for debugging
//         System.out.println(buffer);
     }
     catch( IOException e )
     {
        setBuffer("");
     }
     complete = true;
   }
}
```

Listing 3-5

Continued

Perhaps the most interesting routine, however, is *getResponse*. This function's task is to read a complete response from the SMTP server, and verify it against the expected response codes. If the code matches one of the expected responses, the function returns 0. Otherwise, it returns the response code in question.

The problem with this function is that it should only wait for a certain amount of time before it gives up. One answer would be to check the socket for available characters until a certain time expires. However, this would make your program run continuously while waiting, consuming system resources. Instead, I decided to make a small private class, *TimeoutRead*, which extends *Thread*. This object will execute in a separate thread, which calls *readLine* on the socket. Once the code finds the end of the response, the thread ends.

The main program, then, must wait for the thread to end, or a timeout period to expire. That's exactly what the *Thread.join* function does. The main code calls *join*, passing it the number of milliseconds it is willing to wait. When *join* returns, either the thread completed, or the timeout expired.

The only problem with this scheme is that the thread doesn't end just because the timeout expires. Although the *Thread* object has *stop* and *suspend* methods, they are deprecated (that is, included for backward compatibility, but not recommended for use). Besides, there are no guarantees that these methods would interrupt *readLine*, anyway. If the thread is still running, it may interfere with the program shutting down (not to mention that it will waste system resources). Calling *System.exit* will end the thread, but that might not be appropriate for all programs.

That's why it is important that *sendMail* closes the socket under all conditions. When the socket closes, the *readLine* function throws an exception and this will end the thread.

The *SMTP* class also has a *main* function that you can use to test the class. It takes an e-mail address as a command line argument and sends a short test message to that address. Of course, if you are using the class as part of a larger system, nothing will ever call *main* and you could safely delete it.

Just Enough javadoc

To make it easy for others to use the SMTP class, I decided to create javadoc pages for it. You've certainly seen javadoc documentation before—these are the Web pages that show the class hierarchy and details for your objects. There are many advanced ways to use javadoc, but for simple projects, you really only need to know a few things.

All javadoc comments start with /** and precede some public element (like a class or field). If you put such a comment before a class definition, or a field, or a method, javadoc will pick up the contents of that comment and place it with that item. You can use HTML markup if you like to enhance the appearance of the generated Web page.

Inside these comments you can use special commands that begin with an @ character. For example, *@author* specifies the author's name. There are only a few of these commands you really need for most projects (see Table 3-3). You can also find examples of how these are used in the accompanying listings.

E-Mail Wrap-Up

The focus of this object is the SMTP protocol, not the e-mail message itself. You'd still need to do a bit of work to send, for example, MIME (Multipurpose Internet Mail Extensions)-formatted messages. SMTP doesn't really care what you send, so it should be possible to extend the *SMTP* object to send binary mail and attachments, as well as do other fancy e-mail tricks.

If you decide to pursue higher-level mail functions, you can start by reading through the RFCs. Many of them show extensions to the e-mail format and SMTP. In particular, RFC1341 is probably a good place to start (see Online Resources at the end of this chapter).

TABLE 3-3

Basic Commands for javadoc Allow You to Easily Generate HTML Documentation from Your Source Code

Command	Meaning
@author	The author's name (should use more than one command if more than one author exists)
@version	The version number
@param	Identifies a method's parameters (if any)
@returns	Specifies a method's return value (if any)
@see	Refer to another item (for example, a class or a member of a class)
@exception	Identifies an exception the item throws

Serial Port Events

Listing 3-6 shows another Basic Stamp program. This time, the program waits for a serial data request (from 0 to 3) followed by a nonnumeric character (for example, an equal sign). The Java program that interfaces for this device has to send a request periodically, and then wait for a response.

Sending data via the serial port isn't much more difficult than receiving it. You'll get an *OutputStream* instead of an *InputStream*. You can use it directly, or you can attach it to other I/O classes to perform buffering or formatting.

However, to make things a bit more interesting, I made this particular program handle incoming characters using events. This allows the main program to do something other than wait (potentially forever) for data to arrive from the Basic Stamp.

In the last program, I took advantage of the fact that the serial port, by default, uses 9600 baud, 8 data bits, and no parity. In this program (Listing 3-7), I explicitly set these values (always a good idea) using *setSerialPortParams*.

Superficially, this new program resembles the original serial port program. However, this program must create an object (since an object must implement the event listener interface). The *main* routine simply opens the port and sets the baud rate. It then instantiates the object and calls *go*.

The *go* method retrieves the serial port output stream and also registers the current object as a serial port event listener using *addEventListener*. Of

Listing 3-6
Basic Stamp Temperature Transmitter

```
pot con 7        ' sensor input
sport con 16     ' serial port
baud con 84      ' baud rate constant (84 is 9600 8 N 1)
counts var word
cmd var byte
top:
  serin sport,baud,[dec cmd]
  high pot+cmd ' discharge pot
  pause 1
  rctime pot+cmd,1,counts
  serout sport,baud,["*",hex4 counts,13,10]
  goto top
```

```
import javax.comm.*;
import java.util.*;
import java.io.*;
public class EventSerial implements SerialPortEventListener
{
 SerialPort port;
 EventSerial(SerialPort _port)
 {
    port=_port;
 }
 public void go() throws Exception
 {
    OutputStream ostr=port.getOutputStream();
    String cmd = "0=";  // read channel 0
    port.addEventListener(this);
    port.notifyOnDataAvailable(true);
    port.enableReceiveTimeout(10);  // 10ms timeout
    if (!port.isReceiveTimeoutEnabled())
       System.out.println("Warning: port does not support timeout!");
    // main routine
   while (true)
       {
             byte[] bytes=cmd.getBytes();
             ostr.write(bytes);
             Thread.sleep(10000);
       }
 }
 public void serialEvent(SerialPortEvent ev)
 {
   try
   {
     InputStream str=port.getInputStream();
     while (str.available()!=0)
        {
                int inbyte=str.read();
```

Listing 3-7
Serial Port Program That Creates an Object

```
                char c=(char)inbyte;
                System.out.print(c);
        }
    }
    catch (Exception e)
        {
            System.out.println("Exception: "+e.getMessage());
        }
    }
    public static void main(String args[]) throws Exception
    {
        CommPortIdentifier cid;
        cid=CommPortIdentifier.getPortIdentifier(args[0]);
        SerialPort lport=(SerialPort)cid.open("SerialDemo",2000);
        lport.setSerialPortParams(9600,
            SerialPort.DATABITS_8,
            SerialPort.STOPBITS_1,SerialPort.PARITY_NONE);
        new EventSerial(lport).go();
    }
}
```

Listing 3-7
Continued

course, the program doesn't want all possible events, so it calls *notifyOn-DataAvailable* so that Java will call the class listener method when data arrives at the port. Finally, the program sets a receive timeout of 10 ms. Not all javax.comm implementations support this, but if it is successful, any read will timeout after 10 ms with no characters arriving.

With all the initial housekeeping out of the way, the *go* method enters a perpetual *while* loop. Using the raw *OutputStream,* the program outputs the command to read channel 0. Presumably, the program would then do something else, although in this case, it simply sleeps for 10 s.

The Stamp always echos whatever you send to it (this is a byproduct of its hardware design, which derives its transmit voltage from the receive signal). Therefore, the characters sent to the Stamp will also appear at the program's serial input. A real program would need to strip these out before processing. In this case, the command marks the channel read, which is acceptable.

The *serialEvent* method executes when data is available at the serial port (since no other events are active). The method finds the input stream object and reads bytes from it until the stream reports there are no more bytes to read. In this example, the program simply prints all the bytes out to the console. However, it wouldn't be difficult to build up a complete line of data and then send it via e-mail using the *SMTP* class. Try your hand at integrating these two classes if you like.

SUMMARY

In later chapters, you'll find that you can make a small microcontroller connect directly to your network (or the Internet) via a modem or a network adapter. However, there are still many benefits to using a PC gateway to connect devices to the Internet as this chapter has done.

Many older devices will only have serial ports. While you might adapt them to dialing a modem, establishing network protocols over a modem is not for the faint of heart (nor for the low of memory or processing speed). While you could redesign everything to use a specialized network interface, sometimes a PC (or a small form factor embedded PC) is more cost-effective. This is especially true if you can use PCs with multiport serial boards to handle multiple devices. One PC can effectively act as a gateway for an entire shop floor of embedded controllers.

While the PC is not a classic embedded controller, it has the advantage of being nearly universal. What's more, there are abundant tools available, and cross development on your desktop PC is easy and painless. PC/104 form factor boards and rack-mounted PCs can drive your development costs down. In many cases, a PC is already part of the system, and all that is required is a bit of software to make the jump to the Internet.

Of course, one answer does not fit all. If you are producing devices in volume, PCs—even inexpensive PCs made for control applications—are too expensive. In hostile environments, hardened PCs become even more expensive. As tiny as PCs can get today, you still can't cram them into certain spaces.

While a PC gateway isn't for every situation, however, you should certainly consider it before you move to other more development-intensive ideas.

ONLINE RESOURCES ▨ ▨ ▨ ▨ ▨ ▨ ▨ ▨ ▨

PC/104 Resources

http://www.eg3.com/pc-104.htm—Many PC/104 resources including links to vendors and FAQs.

http://www.pc104.org/—The PC/104 consortium.

http://www.controlled.com/pc104faq/—A PC/104 FAQ.

RFCs

http://www.faqs.org/rfcs/rfc821.html. RFC821 describes basic SMTP.

http://www.faqs.org/rfcs/rfc822.html. RFC822 describes the basic e-mail format.

http://www.faqs.org/rfcs/rfc1341.html. If you are interested in MIME e-mail, this is the basic RFC dealing with that (although there are many other RFCs that add enhancements to this specification).

Java Resources

http://java.sun.com/products/jdk/javadoc/index.html/ The javadoc homepage.

http://java.sun.com/products/javacomm/—Sun's home page for javax.comm.

Introducing TINI

When I was a kid, a computer was a giant machine nestled inside an office building tended to by dozens of specialized acolytes. Today, you can hardly find a machine that big anymore. Even big computers are the size of filing cabinets. If you grew up in those days, you think of PCs as "small" and "cheap." After all, in 1972, dynamic random-access memory (RAM) cost about 1¢ per bit ($80,000 a megabyte—if you could afford the space and power required for a megabyte).

The truth is that PCs are small and inexpensive, but for some applications they aren't small and inexpensive enough. Luckily, there is a new breed of microcontroller that can easily replace PC gateways (like the one discussed in Chap. 3). One of these, the Dallas Semiconductor Tiny Internet Interface (TINI), costs less than $50 and is the same size as a standard 72-pin Single Inline Memory Module (SIMM) module. That means the TINI (pronounced like *tiny*) can go places where a PC is impractical.

What do you get with a TINI? The big plus to TINI is that you get the ability to run Java programs. You also get a variety of I/O ports including Ethernet. An Ethernet port and Java together on a $50 board no bigger than a dollar bill folded in half can put almost anything on the Internet.

Using the TINI is a bit strange. You do your development on a PC or other Java workstation. When you are ready to download your program, you don't burn an EEPROM, or use a special downloader. Instead, you use FTP to send your code to the TINI board. Then you can telnet into the board and use the command shell (slush) to run and test your program. If you don't want to run the shell, you can use a special loader program to replace the shell with your program. This might be a good idea for a stand-alone product, but for development, you'll use the shell, FTP, and telnet.

Preparing TINI

The TINI (see Figs. 4-1 and 4-2) needs—at least—a SIMM socket, power, and I/O connections to run. There are several carrier boards available from Dallas Semiconductor and other companies that will help you get started developing. These carriers have voltage regulators, RS232 jacks, and Ethernet jacks, so they are simple to get up and running. Keep in mind, the TINI itself has all the RS232 and Ethernet hardware—the carrier simply provides the physical connection points.

Figure 4-1
The TINI (Front View)

Flash ROM
(Java™ Runtime Environment)   Microcontroller

Ethernet controller   72-pin SIMM   SRAM   RS232

Figure 4-2
The TINI from the
Rear

IEEE Ethernet address
                    Real time clock

SRAM nonvolatizer

1-wire net driver   Lithium backup

Before you can start using the TINI, you'll likely want to load the latest version of the TINI's firmware and the slush command shell. Loading this basic software is almost certainly the hardest part of using TINI. Luckily, you only have to do it once.

The first thing you need is Java. The TINI documentation will tell you which version of Java Dallas Semiconductor supports (I used 1.2.2 for the examples in this book). You need the Java runtime so you can run the utility programs that set up the TINI. However, you'll also need to compile programs, so you might as well get the whole JDK from Sun (see Chap. 2 for more information about downloading the JDK from Sun). In addition, you need the Java extensions that use the serial port (javax.comm). These are not available for all platforms, so this will restrict your choice of development platforms somewhat.

Once you have the TINI in its carrier, you'll need to supply at least power and an RS232 cable to the development workstation (in other words, your PC). Next, you'll need to run the JavaKit program that Dallas Semiconductor supplies.

There are several things you'll need to set up before you can run JavaKit. First, you must have the Dallas Semiconductor software on the same disk drive as the javax.comm library files (this is a bug in the Sun library). You don't need to use the same directory, but it must be on the same disk. Suppose your system is set up like this:

```
Java SDK—d:\jdk1.2.2
Dallas TINI—d:\TINI
```

Your CLASSPATH variable should already be set to *d:\jdk1.2.2\lib;* or something similar. You still need to add the jar files for javax.comm and the TINI files. You could add these to your CLASSPATH variable, but it is usually just as well to add them as you execute JavaKit. Here is an example command:

```
java -classpath
D:<\\>TINI<\\>bin<\\>tini.jar;D:<\\>jdk1.2.2<\\>lib<\\>comm.jar;%CLASSPA
TH% JavaKit
```

Of course, you can put this line in a batch file, and make it easy to execute the JavaKit with one command.

When you start JavaKit, you should see a screen like the one shown in Fig. 4-3. Of course, as Dallas Semiconductor upgrades its software, it may make slight changes, but you should see all the basic parts of JavaKit on your screen. At this point, you are not communicating with the TINI.

First, you'll need to select the correct COM port. By default, the TINI will use 115K baud, so you don't need to change that setting. Press Open Port to connect to the TINI. Then press the Reset button on the JavaKit screen. This resets the TINI by manipulating the data terminal ready (DTR) line (some carriers may have a jumper to disable this reset function—consult the carrier board's documentation). You should see a prompt something like this:

```
TINI loader 05-15-00 17:45
Copyright (C) 2000 Dallas Semiconductor. All rights reserved.
>
```

You may have special instructions with your TINI that tell you how to update the permanent firmware [the flash read-only memory (ROM)] inside the TINI. Be sure to follow those instructions to the letter. If you mess up the permanent firmware, you'll have to send your TINI back to the factory for repair. Luckily, these firmware updates are infrequent and you may never need to update the firmware anyway.

If this is your first time using the TINI, you should make sure that the latest core software is on board. From JavaKit, select the File | Load

Figure 4-3
JavaKit Screen

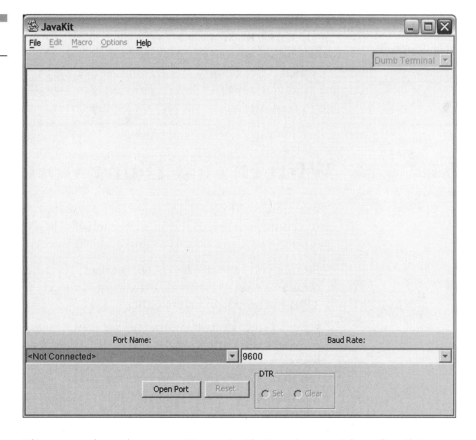

File menu item (or press Control+F). Use the resulting file dialog to select tini.tbin from the TINI's bin directory. This contains the TINI's operating system and other important code. If you don't load this code completely, you can always restart JavaKit and try again, so this isn't nearly as critical as updating the firmware. Again, you don't need to do this every time, just the first time.

The final step is to use JavaKit to load and run a default program. Eventually, this program might be your custom software, but for now you want to use slush—the command shell. Use the File | Load File menu again, but this time select slush.tbin. The slush program looks like a stripped-down Unix shell prompt. It manages the TINI's file system, which is really in memory instead of a disk drive. However, that means you should clear slush's data area. This may change from version to version of slush, but with the version I used, you can issue two commands:

B18
F0

This switches to TINI's bank 18 and fills it with zeros. This gives slush a clean start (which means anything you had stored in a TINI "file" is now lost).

When TINI reboots, it looks for JavaKit on the serial port. If it doesn't find it, it runs the default program. You can also run the default program from the loader prompt by typing *E* and *Enter.*

What If That Didn't Work?

There are several pieces you have to have working correctly before you can communicate with the TINI. Sometimes, it is hard to tell what piece is causing problems. Here is a simple checklist:

1. Be sure you have the Java SDK installed. Try running some sample programs that Sun provides. If they don't work, you can be sure the TINI software won't work either.

2. It is vital that the javax.comm package is working properly. The package has a BlackBox sample included. Build it and run it (using the instructions provided by Sun) to make sure everything is working. If you don't see the COM ports you know are on your machine, something is wrong. Don't forget, many versions of javax.comm want your program to reside on the same disk as the javax.comm jar file.

3. Make sure tiniclasses.jar is not in your CLASSPATH variable. This jar file contains classes that run on the TINI, not on the workstation.

4. With JDK 1.1 you must add swingall.jar to your CLASSPATH.

5. Make sure the serial port on your desktop computer is not in use by other software (modems, hand-held computer hot synch programs, etc.)

6. Be aware that many PCs will not allow you to use COM1 and COM3 simultaneously (the same restriction applies to COM2 and COM4). So if you are using COM1, COM3 may not be available.

7. Check to make sure you are using the correct cable for your carrier board. The standard boards from Dallas Semiconductor need a straight cable (as opposed to a cross cable). A cable that connects two PCs will not work. If you are using adapters, make sure they connect all the lines, since JavaKit uses more than just three wires to communicate.

Logging In

By now you should have slush running in the JavaKit terminal window. The prompt should read:

```
Hit any key to login.
```

Naturally, press any key and log in as user root with password *tini* (there is also an account with the name *guest* with the password also *guest*). You'll find yourself at a command prompt that is similar (but not as powerful as) a Unix shell prompt. Type *ls* and you'll see any files in the root directory (at first, this will be just an *etc* directory).

Using slush

Getting around in slush is easy—especially if you are comfortable with Unix or Linux. You can use *cd* to change directories (enter *cd etc,* for example, to see what's in the *etc* directory using the *ls* command). Use *cd* to go back to the parent directory.

You can also type *help* to see a list of valid commands. For now, you really don't need any of these commands, because you still aren't quite ready to download and run a program. Why not? The Ethernet adapter is still not set up properly.

On the Net

Before you get started with TINI, you need to configure its Ethernet port. You do this with slush's *ipconfig* command. There are two ways to obtain an IP address. First, you can ask your network administrator for an IP address. You can't just pick random IP addresses—each network has particular rules about addresses and, if you pick one that doesn't match your network, the TINI won't work. You might also accidentally pick an address that is in use, which would not be good.

In general, you need to know several things about your network (and the *ipconfig* options they correspond to):

1. Your IP address (–a)

2. Your subnet mask (–m)

3. Your domain name (–n)

4. Your gateway (–g)

5. One or two domain name servers (DNSs) (–p and –s)

6. Your SMTP gateway for sending mail (–h)

You may not need all of these, depending on what you want to do. For example, if you aren't sending e-mail, you won't need the –h option. Suppose your network administrator tells you to use 192.168.0.200 for an IP address. The subnet mask is 255.255.255.0 (which will be the same throughout your network). You can enter the following slush command:

```
ipconfig -a 192.168.0.200 -m 255.255.255.0
```

This will set the TINI's address, which will be remembered by the system until you change it or completely wipe out the slush memory file system. If you use the –f option, *ipconfig* will not bother you with warning messages. Without this option, *ipconfig* will warn you that you will shut down any connected users when you change addresses. Since the TINI has no address until you set it, that's not a real problem.

Many networks now use a Dynamic Host Control Protocol (DHCP) server to assign IP addresses dynamically. If you have one of these, your TINI can automatically get an IP address from the server. Of course, the TINI may not always get the same IP address, which can be a problem in some cases. The command to use DHCP is

```
ipconfig -d
```

Again, you can use –f to ignore warning messages. The version of TINI I used had a problem using a Windows NT DHCP server, but presumably that has been fixed by now.

By default, the TINI's host name is TINI. You can change this using slush's *hostname* command. If you use DHCP, you'll probably be able to refer to the TINI by its host name when communicating with it from a workstation. If you aren't using DHCP, you'll have to use the IP address unless you know how to add the name to your DNS server or the workstation's host database (for Unix, try /etc/hosts; for Windows 98, look in c:\windows\hosts; for Windows NT or Windows 2000, you'll need the c:\windows\system32\drivers\etc\hosts file).

In addition to network configuration parameters, you also need an RJ45 connector that connects to your network (known as a *10baseT con-*

nector). One end of this wire plugs into your TINI and the other end connects to a network hub (often stuffed away in an equipment closet, or overhead). This hub, in turn, connects to other computers using the same sort of connectors, and it may also connect to your network's backbone using a different kind of cable and connector. Assuming your network has a connection to the Internet, this simple wire can be a gateway to the entire world.

Once the network setup is complete, you should be able to ping the TINI at its IP address. If you are using DHCP, you may have to use JavaKit to log into slush and type *ipconfig* to see what address the TINI drew from the DHCP server.

If you can ping the TINI, everything should be fine. You should also be able to log into the TINI (using JavaKit) and ping your workstation. If that works, try using an FTP program or telnet program to log into the TINI. The TINI's FTP server is somewhat simplistic, so some advanced FTP programs (like the popular ncFTP and many graphical FTP programs, for example) will report errors. That's because they are sending the TINI commands that TINI doesn't support. However, most basic FTP programs will work fine.

No Network?

What happens if you don't have a network? Or if your network administrator won't let you have an IP address? Then you'll have to build your own mini network. Don't worry—it isn't very hard, and it also isn't very expensive.

Most TINI boards supply an RJ45 jack that accepts 10baseT Ethernet cable. This is the sort of cable that looks like a large phone cable. You have two choices. You can buy a hub (hubs with three or four ports are not very expensive) and an Ethernet card for your computer. If you don't want a hub, you may be able to use a special cable known as a *cross cable* to connect the TINI directly to your Ethernet card. With a cross cable, you can only connect two devices together—you can't add more computers, or anything else like a digital subscriber line (DSL) modem. Also, most Ethernet card manufacturers will not support the use of a cross cable. Still, for a quick, inexpensive setup, a cross cable is a good bet.

If you are setting up your own network, you can assign IP addresses as you see fit. You can even set up DHCP if you like. Of course, the details for this will vary wildly from operating system to operating system.

When you set up IP addresses, you should know that the addresses are in groups (for example, 10.1.1.1 to 10.255.255.255). The part that doesn't change (the 10 in this example) is the network number. The rest of the numbers uniquely identify each machine on the network. The network mask tells you what part of the number doesn't change. For example, for the range I just mentioned, the network mask is 255.0.0.0. If you aren't connected to another network, you could—in theory—select any address range and mask that you like. However, if there is any chance you might be connected—even temporarily—to a public network (like the Internet), you should stick with IP addresses that are reserved for small private networks. In particular, 10.1.1.1–10.255.255.255 (with a net mask of 255.0.0.0) and 192.168.0.1–192.168.0.255 (net mask of 255.255.255.0) are especially for small private networks like this. Public network routers and gateways won't try to pass traffic for these addresses.

Building a TINI Program

Building a TINI program is a two-step process. First, you compile your program using a Java compiler. However, you provide the compiler with the TINI's class library instead of the default library it would usually use. This generates, as usual, a class file. If you were developing for the PC, you could execute these class files directly. However, this isn't true with the TINI. To help make execution more efficient, the TINI requires a linking step to bundle your classes into a single file that the TINI can execute. You can elect to link the program as a tini file that you can run from slush, or as a tbin file that you can load (in place of slush) using the loader. For now, assume you'll make a tini file.

Armed with the tini file, you can use FTP to send the file to the TINI. Then you can telnet into the TINI and run your program.

Whenever I am trying a new microprocessor, I usually just try to get an LED to flash—that should be simple enough and it lets me work the kinks out of the development process.

Listing 4-1 shows a very simple Java program that flashes the LED built into the TINI. You'll notice that the class (*LED*) has only two members. The *bp* field holds a *BitPort* object that represents the LED (on port 3, bit 5). The *main* function, of course, makes the LED blink from within a *for* loop. Notice that I marked *main* so that it can throw any sort of exception. This allows the program to ignore exceptions (like the *InterruptedException* that *Thread.Sleep* might throw).

```java
import com.dalsemi.system.BitPort;
public class LED {
  static BitPort bp=new BitPort(BitPort.Port3Bit5);
  public static void main(String args[]) throws Exception {
     int i;
    System.out.println("Here we go!");
    for (i=0;i<1000;i++) {
        bp.clear();
        Thread.sleep(250);
        bp.set();
        Thread.sleep(250);
    }
  }
}
```

Once you have the program saved (remember, you must use LED.java as the file name), you can compile the program. If you are using the Java SDK, you'll use the following command on the PC:

```
javac -bootclasspath c:<\>tini<\>bin<\>tiniclasses.jar LED.java
```

Of course, you need to replace c:\tini with the correct directory for your setup. This will create the LED.class file. Next, you'll need to convert the file to a tini file. That takes this command line:

```
java TINIConvertor -f LED.class -o LED.tini -d
c:<\>tini<\>bin<\>tini.db
```

That's it! You should now have a file named LED.tini that will run on the TINI hardware. The trick is to now get it over to the TINI and start it running.

Testing the Program

The resulting tini file must be placed in the TINI's file system so that you can test it. The easiest way to do this is with an FTP program. Start your FTP program using the TINI's host name or IP address. Then, log in using the user name root and the password *tini.* The details will depend on what FTP program you are using, but, in general, you can issue a PUT command (as in, PUT LEDtini) to transfer the file to the TINI. Once the file is in place, you'll want to start it executing.

To start the program, you'll need to log in to slush. You could use JavaKit and the serial port as before, but usually you won't keep the serial cable connected for very long after you do the initial setup. Instead, you'll probably want to use telnet. Most operating systems have a default telnet client, and you can find plenty of freeware telnet clients on the Web.

Once again, you can use the root account with the password *tini.* Once you see the slush prompt, you can execute an *ls* command to make sure you can find your LEDtini file. If you created a subdirectory and put your file there, you'll need to use the *cd* command to change to that directory. If you followed the above directions, however, your file should be in the root directory, which is fine.

Once you've located the file, enter the following command:

```
java LED.tini &
```

Case matters, so if your file is named *led.tini* instead of *LED.tini,* use that name instead. The *&* tells the TINI to run the program and return to the slush prompt. Since this program will never stop running, that's important. Without the *&*, slush will wait forever for the program to end and you'll need to open another telnet session to stop it.

After you enter the Java command, the LED on the face of the TINI should start to blink. There are actually four LEDs near the top edge of the TINI. The three that are to the right and grouped together are Ethernet status LEDs. The LED that will blink is lined up with these LEDs, but much further to the left, off by itself. You should also see the "Here we go!" message appear in the telnet window.

The LED will merrily blink away forever. If you had a program that had a definite beginning and ending, you wouldn't need the *&* on the command line. It would be OK to let slush wait for the program to complete.

How do you stop the program? You need to find the process ID that corresponds to the running program. You do this with the PS command. Your results may differ, but here's what my PS command reported:

```
3 processes
1: Java GC
2: init
6: LED.tini
```

Notice the number *6* in front of the *LED.tini* process? That's the program's process ID. You can issue a kill command to kill process *6:*

```
kill 6
```

This will end the test program, and leave the LED in whatever state your program last set it before it terminated.

You've successfully downloaded a program using FTP, run it with the Java command, checked its status with *ps*, and stopped it with *kill*. This program doesn't generate a lot of console output, so it was no problem to let it simply write to the telnet window. However, if your program generates a lot of output, you might not want it cluttering up the terminal window. In this case, you could issue the command

```
java LED.tini &  > txt.out
```

You can use any name you want for *txt.out*—the output will appear in this file. Notice that there is a space between the > and the file name. This space is not optional. If you omit it, slush will complain that you did not provide a file name for redirection.

A New Default Program

Running programs with slush is fine for debugging—in fact, you'll probably prefer it since it gives you a nice way to start and stop programs, read console output, and generally control the TINI. However, once your code is running, you might not want to use slush.

TINI allows you to create programs that will load in place of slush. The build process for these files is practically identical to the one just described. The only difference is that the TINICONVERTER program requires a -l (that is, a lower case L) option:

```
java TINIConvertor -l -f LED.class -o LED.tbin -d
c:<\>tini<\>bin<\>tini.db
```

Also, by convention, the output file name ends with *tbin* instead of *tini*. Once you have the tbin file, you can't download it using FTP. Instead, you'll need to start JavaKit and load your file just like you loaded slush previously. This will wipe out slush, so make sure you have any file on the TINI that you want backed up to your PC. Just use File |

Load File from the JavaKit menu and pick your tbin file from the dialog that appears. Now when the TINI starts, it will run your LED flashing code. You can also issue an E command to the loader to make it start the program.

Keep in mind that, without slush, you can't communicate with the TINI other than via your own program. If you want to change anything, you'll have to use JavaKit to invoke the loader. If you want to use slush again, you'll need to load it from its tbin file. You should also plan on resetting the slush heap since your program may use it. That means you'll lose everything in the TINI's file system—so back it all up before you load your own programs!

If you just want the TINI to start a program automatically, but you don't want to replace slush, you can do that too. Simply put the slush commands you want to execute on startup in the TINI's /etc/.startup file. Be sure to download the existing file, as it contains some network initialization you'll want to keep. Once you edit the file, you can save it back (using FTP) and restart the TINI to test it.

Serial Ports

Freeing the Serial Port

Consider the program in Listing 4-2. This program opens the TINI's serial0 port (the same RS232 port that you use with JavaKit). It then sends out 100 lines of output at 9600 baud.

This poses several problems. First, slush is listening on the serial port for log-ins. Second, the port's hardware is wired so that DTR causes a processor reset. There are several possible solutions for each problem. First, if you are using telnet and FTP to talk to TINI, you can turn off slush's serial log-in. You can do this by issuing the slush command

```
stopserver -s
```

You can also use the *downserver* command, which is a synonym for *stopserver*. Once you stop the serial server, you can use the port for your own programs. That doesn't solve the reset problem. If you have a breakout box, you can simply disconnect the DTR pin (pin 4 on the DB9). You could also build a special cable with pin 4 missing. Another solution is to

Editing on the TINI

In general, the only way to edit a file on the TINI is to use FTP to download it to your host computer, edit it, and then reload it using FTP again. Of course, many times, your PC already will have the exact file (like a tini program file, for example) so you don't have to download it; you simply modify it and copy it to the TINI.

However, there are times when you might want to make a quick change to a property or a data file that resides on the TINI. At http://www.smartsc.com/tini/TEd/ you can find a simple text editor for the TINI. However, be warned that it is a part of slush. That means you have to replace slush (using JavaKit) with the slush available on this Web site. You can also get the source code and build it into your own copy of slush (which is good if your version of slush does not match the author's). However, in either case, you need to use JavaKit to reload the custom version of slush.

```java
import java.io.*;
import java.net.*;
import javax.comm.*;
public class SerialSender
 {
 SerialPort rs232;
 public static void main(String args[])
  {
  new SerialSender().go();
  }
 void procError(String s) // Note: Tini Exceptions don't return messages
  {
  System.out.println("Error");
  System.out.println(s);
  System.exit(1);
  }
```

Listing 4-2
Sending Strings Over the Serial Port.

```
public void go(String args[])
  {
    try
      {
// Use this code to open the RS232 port
    CommPortIdentifier cpi;
      cpi=CommPortIdentifier.getPortIdentifier("serial0");
      if (cpi==null) procError("Can't find SERIAL0");
      rs232 = (SerialPort)cpi.open("SSend",1000);
    rs232.setSerialPortParams(9600,SerialPort.DATABITS_8,
      SerialPort.STOPBITS_1,SerialPort.PARITY_NONE);
    is=rs232.getInputStream();
    }
    catch (Exception e)
    {
    procError("Can't open RS232 port " );
    }
int i;
PrintStream os=null;
try {
 os=new PrintStream(rs232.getOutputStream());
} catch (IOException e) { System.exit(1); }
for (i=100;i>100;i--) os.println(i+ " bottles of beer on the wall");
}
}
```

Listing 4-2
Continued

use JavaKit and make sure the DTR clear radio button is checked. This will prevent the processor from resetting.

If you find you need the serial port for your programs often, you may want to permanently stop slush from looking for serial log-ins. Simply edit the .startup file (in the /etc directory) and change the line that reads

```
setenv SerialServer enable
```

Naturally, you'll change the *enable* to *disable*. The change won't take effect until you restart the TINI. If you do this, and for some reason you can't telnet into the TINI, you'll need to clear the slush heap to get serial log-ins working again. Of course, if you can telnet into the TINI, you could always manually start the serial server if you needed to do so.

More Practical Serial

The TINI has lots of potential. With its multitude of interfaces, it is possible to use it as an intelligent bridge between practically any device and any network (including the Internet). Suppose you want to measure several process variables in a manufacturing plant. Since there is a Web server that supports servlets available (see Chap. 5), you could simply collect the data using the TINI and make the parameters available as part of a Web page. Anyone on the network could monitor the plant's operation using an ordinary Web browser. The Java code could perform engineering unit conversion or averaging, or store historical data.

Of course, you might want another program to read the data instead of a Web browser. That's even easier, since Java makes working with sockets a snap. Since many devices already use RS232, I decided to build a TINI RS232 to Ethernet bridge. The bridge accepts input from the RS232 port and sends it to a socket on a remote computer. A companion Java program running on the remote computer simply displays the data, but in real life, it might display the results graphically, or save the values in a file.

RS232 Woes

The TINI technically has support for four serial ports. However, one of them is a time to live (TTL) serial port and doesn't have full support in the current release of the firmware. Two other ports require external universal asynchronous receiver transmitters (UARTs). That leaves the main serial port as the only functional true RS232 port on the board without adding more hardware.

Unfortunately, slush and the system console use this serial port. You must shut down slush's serial server (using downserver -s -d) if you want to use the serial port from your own program. You can also modify the /etc/.startup file to prevent the serial server from loading. If you replace slush with your own program, you can use the serial port freely.

Dallas Semiconductor supplies a custom object that allows you to use the serial port (the source code for slush is a good example of how to use it). However, recent versions support the standard javax.comm package, so the custom routines are now deprecated.

With some experimentation, you'll find that the serial port doesn't support all possible baud rates, nor does it support flow control. I couldn't find a single place where the capabilities were documented, but your program will throw an exception when you try to use something that isn't there.

Inside the Project

Listing 4-3 shows the *SerialSender* class, a Java program that requires two arguments. The first argument is the IP (or host name) of the computer that should receive the data. The second argument is the port number that computer is using. Most of the action occurs in the *go* subroutine.

```
import java.io.*;
import java.net.*;
import javax.comm.*;
public class SerialSender
 {
 SerialPort rs232;
 Socket sock;
 OutputStreamWriter os;
 InputStream is;
 String msg;
 public static void main(String args[])
  {
  if (args.length!=2)
    {
    System.out.println("Usage: SerialSender host port");
    System.exit(9);
    }
  new SerialSender().go(args);
  }
```

Listing 4-3
Sending from a Network Port

```
    void procError(String s) // Note: Tini Exceptions don't return messages
    {
    System.out.println("Error");
    System.out.println(s);
    System.exit(1);
    }
    public void go(String args[])
      {
      try
        {
          System.out.println("Connecting to " + args[0] +":"+ args[1]);
          sock=new Socket(args[0],Integer.parseInt(args[1]));
        }
      catch (Exception e)
        {
        procError("Can't open Socket");
        }
      try
        {
        os=new OutputStreamWriter(sock.getOutputStream());
        os.write("Start\r\n",0,7);
        os.flush();
        }
      catch (Exception e)
        {
        procError("Can't open stream");
        }
      try
          {
// Use this code to open the RS232 port
//    CommPortIdentifier cpi;
//    cpi=CommPortIdentifier.getPortIdentifier("serial0");
//    if (cpi==null) procError("Can't find SERIAL0");
//    rs232 = (SerialPort)cpi.open("SSend",1000);
//    rs232.setSerialPortParams(9600,SerialPort.DATABITS_8,
```

Listing 4-3

Continued

```
//      SerialPort.STOPBITS_1,SerialPort.PARITY_NONE);
//   is=rs232.getInputStream();
// Or... Just use stdin
     is=System.in;
   }
  catch (Exception e)
   {
   procError("Can't open RS232 port " +msg);
   }
  while (true)
   {
   int c;
   try
     {
     c=is.read();
     }
   catch (Exception e)
     {
     continue; // ignore rs232 exceptions
     }
   try
     {
       if (c==13) c=10;
     os.write(c);
       if (c==10) os.flush();
     }
   catch (Exception e)
     {
     procError("Write error");
     }
   }
  }
}
```

Listing 4-3
Continued

First, the program opens a socket. The program then writes the string *Start* to the socket as a debugging aid. Note that the code flushes the socket. This way if the RS232 code goes astray, you'll still see the debugging message. Without the flush, the string could be waiting in limbo after the code hangs. This gives you the incorrect impression that the socket code did not work, when in fact it is the RS232 code that is the culprit.

The javax.comm package provides a standard way to use communication ports under Java. The basic sequence of events that you'll use to open an input stream on a serial port is the following:

1. Create a *CommPortIdentifier* object that names the serial port (for the TINI this is *serial0*) by calling the static method *CommPortIdentifier.getPortIdentifier.*

2. Use the object's open method to create a *SerialPort* object.

3. Use *setSerialPortParams* to set the *SerialPort's* baud rate and other parameters.

4. Call *getInputStream* to return an *InputStream* object that refers to the port.

This *InputStream* object is like any other input stream. In fact, for debugging purposes, you might want to replace the serial port code with a single line:

```
is=System.in;
```

This allows you to use the standard input stream as a source for input (so you can enter characters from the slush terminal or a telnet session).

After both the port and the socket are ready, a simple *while* loop reads characters from the port and sends them to the socket. On receipt of an end of line character, the program flushes the socket so that the remote computer receives frequent updates.

The Other Side

Listing 4-4 shows the Java program for the host computer. This is just an ordinary Java program—you don't need any special TINI classes or conversion to run this program. It requires one argument—a port number.

Instead of an ordinary socket, the host program uses a *ServerSocket.* This specialized class handles all of the details required to wait for client

Listing 4-4
The Serial Receiver

```java
import java.io.*;
import java.net.*;
public class SerialRcv
{
ServerSocket ssock;
BufferedReader br;
Socket sock;
public static void main(String args[])
  {
  if (args.length!=1)
    {
    System.out.println("Usage: SerialRcv port");
    System.exit(1);
    }
  new SerialRcv().go(args);
  }
public void go(String args[])
  {
  try
    {
    ssock=new ServerSocket(Integer.parseInt(args[0]));
    System.out.println("Listening on port " + args[0]);
    sock=ssock.accept();
    ssock.close(); // quit listening
    System.out.println("Status: Connection established");
    br=new BufferedReader(
       new InputStreamReader(
         sock.getInputStream()));
    }
  catch (Exception e)
    {
    System.out.println("Can't open or connect server socket");
    System.exit(1);
    }
  while (true)
    {
    try
```

Listing 4-4
Continued

```
      {
      String s=br.readLine();
         if (s==null) break;
         if (s.length()==0) continue;
      System.out.println(s         );
      }
    catch (Exception e)
      {
      System.out.println("Read error");
      System.exit(2);
      }
    }
  }
}
```

requests. Since the program only handles one client at a time, the program closes the socket after the call to accept (which returns when a client connects).

Once the socket is ready, it is a simple matter to make a *BufferedReader* and read the incoming data a line at a time. The example in Listing 4-4 simply echos the data to the console output, but the same principle would apply if you wanted to do more sophisticated processing.

About One Wire

The TINI has numerous expansion options. The board supports several serial ports, parallel I/O, and several other expansion ports—however, many carrier boards don't support all of these options, which means you'd have to build custom hardware to take advantage of them.

One area where the TINI is very capable is in the One Wire interface. This isn't surprising since Dallas Semiconductors—the same company that makes TINI—created One Wire. As an example, consider the One Wire DS2406 device. Each of these devices—using a single wire—can manage two switches or other sensors. Figure 4-4 shows a simple board that connects to the TINI and manages eight switches (using four DS2406 devices) and also shows the status of the lines with LEDs. That means

Figure 4-4
A TINI Interface Board

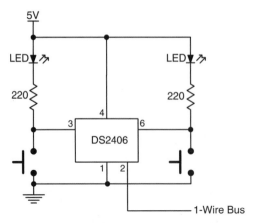

5V

LED

220

4
3   6
DS2406
1   2

1-Wire Bus

(Repeat circuit 4 times, connecting each section
to the 1-Wire Bus)

you can see when the switches are closed because the LEDs will turn on.
Also, The TINI can set the status of the LEDs by latching an input (this
is a feature of the DS2406).

Consider the simple code in Listing 4-5. This program acts like a flip-
flop. If you close the switch on the first DS2406, the LED on the last
switch turns on. If you close the switch of the second DS2406, the LED
on the last switch turns off. These switches might represent some real-
world sensor like a liquid level sensor.

Soon, you'll create more code to Internet-enable this flip-flop. First,
however, try building this code and you'll see there are a few peculiarities to the build process that didn't come up before.

You'll notice that the program starts a new thread of execution in the
iAmAlive method. This thread blinks the light on the TINI so you can
tell the program is running. Since it is in a separate thread, it doesn't
affect the execution of the program's code at all. However, I used an
anonymous class to create the thread. This causes the Java compiler to
generate a file named Switch$1.class. That's no problem, but you must
include this file when you run TINIConvertor. Otherwise, you'll have a
reference that is not satisfied and the build will fail.

Using the *iAmAlive* method is often useful to make sure your program is
functioning during development. If you crash the TINI, your FTP and telnet sessions may not work. Also, if you have to load the program without
slush, the LED can give you an indication that your program is operating.

Another issue arises if you are not using the latest version of the
TINI firmware. Between version 1.01 and 1.02, Dallas Semiconductor

```
import java.util.*;
import com.dalsemi.onewire.OneWireAccessProvider;
import com.dalsemi.onewire.adapter.DSPortAdapter;
import com.dalsemi.onewire.container.OneWireContainer12;
import com.dalsemi.system.BitPort;
public class Switch {
  OneWireContainer12[] buttons = new OneWireContainer12[4];
  byte[][] ary = new byte[4][];
  public static void pause(int ms) {
      try { Thread.sleep(ms); }
      catch (InterruptedException e) { }
  }
  public static void iAmAlive() {
      // blink LED on TINI board
    (new Thread() {
        BitPort bp=new BitPort(BitPort.Port3Bit5);
          public void run() {
            while (true) {
                bp.clear();
                pause(250);
                    bp.set();
                    pause(250);
                }
              }
          }).start();
}
public static void main(String args[]) throws Exception {
   int i;
System.out.println("Here we go!");
iAmAlive();
   DSPortAdapter pa = OneWireAccessProvider.getDefaultAdapter();
   pa.targetFamily(0x12); // find DS2406
i=0;
   do {
     try {
```

Listing 4-5
Reading Switches

```
      for (Enumeration e=pa.getAllDeviceContainers();e.hasMoreElements();i++) {
          buttons[i]=(OneWireContainer12) e.nextElement();
      }
  }
          catch (Exception e) { System.out.println("Exception");}
} while (i<4);
boolean b=true;
for (i=0;i<4;i++) {
ary[i]=buttons[i].readDevice();
  buttons[i].setLatchState(0,false,false,ary[i]);
buttons[i].writeDevice(ary[i]);
}
while (true)  {
ary[0]=buttons[0].readDevice();
  if (!buttons[0].getLevel(0,ary[0])) {
        setLED(buttons[3],true,ary[3]);
    System.out.println("On");
  }
  ary[1]=buttons[1].readDevice();
    if (!buttons[1].getLevel(0,ary[1])) {
          setLED(buttons[3],false,ary[3]);
      System.out.println("Off");
    }
  }
}
synchronized static void setLED(int n,boolean state) throws Exception {
    ary[n]=buttons[n].readDevice();
    buttons[n].setLatchState(0,state,true,ary[n]);
    buttons[n].writeDevice(ary[n]);
  }
}
```

Listing 4-5
Continued

released new One Wire interface classes. However, this poses a problem. Some of the classes you use are in the TINI's firmware that you load using JavaKit. That means that if the names change, you can no longer link to these internal functions. Rather than waste space by duplicating them, Dallas Semiconductor made a set of temporary adapter classes available. These adapters effectively translate some new names into the old ones, without duplicating the old code. If you are using version 1.01, you should consult the TINI documentation to learn more about these classes. Better still, upgrade to the latest version of the TINI firmware.

Notice that to use the One Wire devices, you need two separate classes— an adapter that represents the One Wire port on the TINI (a *DSPortAdapter*) and a container for each device on that adapter (in this case *OneWireContainer12*). Each device has an ID code, and the DS2406's code is 12, which explains the odd-looking name.

The adapter can enumerate all the devices that it controls, and this program uses that feature to find all the DS2406 devices. The container objects allow you to read the device state into an array, examine and alter it, and then write the state back to the device. The program performs these actions in a loop and simply examines the state of each switch, setting the LED state as appropriate.

The first step is to open the adapter:

```
DSPortAdapter pa = OneWireAccessProvider.getDefaultAdapter();
pa.targetFamily(0x12); // find DS2406
```

This opens the adapter and informs it that you only want to work with devices that have a family code of 0x12 (the DS2406). Now you can call *getAllDeviceContainers* to find an *Enumeration* that will list all the DS2406 devices on this adapter. The program stores the containers in an array (*buttons*) and stops when it has found four different devices.

Armed with container objects for the four devices of interest, the program reads the state of each device (*readDevice*) and uses *setLatchState* to turn off all the LEDs. The LEDs are connected to the switches, so the LEDs will light as long as you hold down the switch even with no software control. However, using the latching feature, you can cause the LED to light (or extinguish) even with no button activity. Of course, the LED will always light when you press the button no matter what the software tries to do. Once the program selects the latch state, it calls *writeDevice* to actually update the DS2406.

Next, the program enters an endless loop. Using *readDevice* and *getLevel*, the program samples the first two switches and calls the *setLED* method to set the fourth LED on or off as appropriate. The *setLED* method is not part of the One Wire container. Instead, it is a method of the *Switch* class. It simply wraps the *readDevice, setLatchState,* and *writeDevice* logic required to set the LED's state.

So far, you could do this task with any microcontroller. However, if you want to report the status back to a computer over the Internet, you might have trouble. With the TINI, however, it is easy.

Some Network Examples

If you want to command and control the TINI flip-flop, you have several options. First, you can let the TINI listen to a port that will allow it to accept commands and return results. Second, you can periodically send the current state to a remote computer. You can even send e-mail, like the example in Chap. 3.

Thanks to Java, all of these strategies are straightforward. Consider the program in Listing 4-6. Here, the program starts two threads. The first is

```java
import java.util.*;
import java.net.*;
import java.io.*;
import com.dalsemi.onewire.OneWireAccessProvider;
import com.dalsemi.onewire.adapter.DSPortAdapter;
import com.dalsemi.onewire.container.OneWireContainer12;
import com.dalsemi.system.BitPort;
class ProcessCmd extends Thread {
  Socket sock;
  public ProcessCmd(Socket sock) { this.sock=sock; }
  public void run() {
      try {
          BufferedReader is = new BufferedReader(new InputStreamReader(sock.getInput-
Stream()));
          PrintWriter os=new PrintWriter(new OutputStreamWriter(sock.getOutputStream()),true);
          while (true) {
```

Listing 4-6
A Networked Flip-Flop

```
                    String line=is.readLine();
                    System.out.println("Read: "+line);
                    if (!Switch1.processRequest(line.charAt(0),os)) {
                        sock.close();
                        return;
                    }
                }
            } catch (Exception e) {
                try { if (sock!=null) sock.close(); }
                catch (IOException e0) { }
                System.out.println("Exception while processing cmd: " + e);
            }
        }
    }
class NetServer extends Thread {
    static ServerSocket server=null;
    static int portno;
    NetServer(int port) { portno=port; }
    public void run() {
        if (server==null) {
            try {
                    server=new ServerSocket(portno);
                    while (true) {
                      Socket newsock=server.accept();
                      ProcessCmd cmd=new ProcessCmd(newsock);
                      cmd.start();
                    }
            } catch (Exception e) {
                    System.out.println("Can't start server: " + e);
            }
        }
    }
}
public class Switch1 {
  static boolean state=false;
```

Listing 4-6

Continued

```
static boolean cmdRequest = false;
static boolean stateRequest;
static OneWireContainer12[] buttons = new OneWireContainer12[4];
static byte[][] ary = new byte[4][];
public static void pause(int ms) {
    try { Thread.sleep(ms); }
    catch (InterruptedException e) { }
}
public static void iAmAlive() {
    // blink LED on TINI board
  (new Thread() {
        BitPort bp=new BitPort(BitPort.Port3Bit5);
            public void run() {
              while (true) {
                    bp.clear();
                    pause(250);
                    bp.set();
                    pause(250);
                }
            }
  }).start();
}
synchronized static public boolean processRequest(char c,PrintWriter os) {
 if (c=='*') return false;
 if (c=='1') {
    stateRequest=true;
    cmdRequest=true;
 }
 if (c=='0') {
        stateRequest=false;
        cmdRequest=true;
 }
 if (c=='?') os.println(state?"1":"0");
 return true;
 }
```

Listing 4-6
Continued

```
public static void main(String args[]) throws Exception {
    int i;
  System.out.println("Here we go!");
  iAmAlive();
    NetServer svr = new NetServer(111);
  svr.start();
    DSPortAdapter pa = OneWireAccessProvider.getDefaultAdapter();
    pa.targetFamily(0x12); // find DS2406
  i=0;
    do {
      try {
        for (Enumeration e=pa.getAllDeviceContainers();e.hasMoreElements();i++) {
            buttons[i]=(OneWireContainer12) e.nextElement();
        }
      }
            catch (Exception e) { System.out.println("Exception: "+e);}
    } while (i<4);
    boolean b=true;
    for (i=0;i<4;i++) {
    ary[i]=buttons[i].readDevice();
      buttons[i].setLatchState(0,false,false,ary[i]);
    buttons[i].writeDevice(ary[i]);
    }
    while (true)  {
      if (cmdRequest) {
          cmdRequest=false;
          setLED(2,stateRequest);
      }
    ary[0]=buttons[0].readDevice();
      if (!buttons[0].getLevel(0,ary[0])) {
          setLED(3,true);
          state=true;
      System.out.println("On");
      }
    ary[1]=buttons[1].readDevice();
```

Listing 4-6

Continued

```
        if (!buttons[1].getLevel(0,ary[1])) {
              setLED(3,false);
              state=false;
          System.out.println("Off");
        }
      }
  }
synchronized static void setLED(int n,boolean state) throws Exception {
    ary[n]=buttons[n].readDevice();
    buttons[n].setLatchState(0,state,true,ary[n]);
    buttons[n].writeDevice(ary[n]);
  }
  }
```

Listing 4-6
Continued

just the *iAmAlive* thread as before. The other thread acts just as the original program did. However, it also spins off another thread (based on the *NetServer* class).

The *NetServer* class listens on port 111 in a loop using *ServerSocket*. Each call to *accept* blocks until a client attempts to connect. When that occurs, *accept* returns an ordinary socket the TINI can use to communicate with the client. However, the *NetServer* thread should continue listening for more connections, so it creates a new thread based on *ProcessCmd*. This thread reads lines from the socket and calls the main object's *processRequest* method.

The *processRequest* method handles four different commands. You can send an ASCII 1 (or 0) to turn on (off) the third LED on the board. You can send a question mark to get the status of the output LED (the fourth LED). Finally, an asterisk causes the TINI to hang up the connection. Of course, if you end the socket connection, the TINI will get an I/O exception, and hang up anyway. That's important, because on the real Internet you may have situations where your connection simply vanishes.

You can build this program just like the previous programs. You'll need to bind Switch1.class, Switch1$1.class, NetServer.class, and Process-Cmd.class into the tini file. When you run the program, it will act just like the original one, except that the program will respond to network requests.

The One Wire interface classes are not intrinsically thread-safe. You can lock the entire adapter using *beginExclusive* and release it with *endExclusive*, but I decided to simply use the One Wire classes from a single thread. The *ProcessCmd* thread only manipulates flags in the *Switch1* class. The main loop within *Switch1* acts on these flags, so only it makes One Wire calls.

Listing 4-7 shows a simple program that sends data to the TINI via the network socket. You can run this program on any Java computer that has a network connection to the TINI (be sure to set the correct address in the socket constructor). By default, the program turns on the LED. However, you can specify any of the legal commands via the command line.

```java
import java.net.*;
import java.io.*;
public class Switch1Test {
    public static void main(String args[]) throws Exception {
        Socket sock=new Socket("TINI",111);
        PrintWriter pw=new PrintWriter(new OutputStreamWriter(sock.getOutputStream()),true);
        String s="1";
        if (args.length!=0) s=args[0];
        System.out.println("Sending: "+ s);
        pw.println(s);
        if (s.charAt(0)=='?') {
            BufferedReader in=new BufferedReader(new InputStreamReader(sock.getInput
              Stream()));
            String line=in.readLine();
            System.out.println("Response: "+line);
        }
        pw.println("*");
        pw.close();
    }
}
```

Listing 4-7
Java Program That Sends Data to the TINI via the Network Socket

Another possibility is to have the TINI actively attempt to inform a remote computer when something changes. This is really just the reverse of Switch1. The TINI, in this case, will have an ordinary socket while the remote program will use *ServerSocket*. You can find Switch2—a program that implements this idea—in Listing 4-8.

```java
import java.util.*;
import java.net.*;
import java.io.*;
import com.dalsemi.onewire.OneWireAccessProvider;
import com.dalsemi.onewire.adapter.DSPortAdapter;
import com.dalsemi.onewire.container.OneWireContainer12;
import com.dalsemi.system.BitPort;
class ProcessCmd extends Thread {
  Socket sock;
  public ProcessCmd(Socket sock) { this.sock=sock; }
  public void run() {
    try {
      BufferedReader is = new BufferedReader(new InputStreamReader(sock.getInputStream()));
      PrintWriter os=new PrintWriter(new OutputStreamWriter(sock.getOutputStream()),true);
      while (true) {
        String line=is.readLine();
        System.out.println("Read: "+line);
        if (!Switch2.processRequest(line.charAt(0),os)) {
          sock.close();
          return;
        }
      }
    } catch (Exception e) {
      try { if (sock!=null) sock.close(); }
      catch (IOException e0) {}
      System.out.println("Exception while processing cmd: " + e);
    }
  }
}
```

Listing 4-8
Switch2 Program

```
    }
class NetServer extends Thread {
  static ServerSocket server=null;
  static int portno;
  NetServer(int port) { portno=port; }
  public void run() {
      if (server==null) {
        try {
            server=new ServerSocket(portno);
            while (true) {
              Socket newsock=server.accept();
              ProcessCmd cmd=new ProcessCmd(newsock);
              cmd.start();
            }
        } catch (Exception e) {
            System.out.println("Can't start server: " + e);
        }
      }
  }
}
public class Switch2 {
  static boolean state=false;
  static boolean cmdRequest = false;
  static boolean stateRequest;
  static OneWireContainer12[] buttons = new OneWireContainer12[4];
  static byte[][] ary = new byte[4][];
  public static void pause(int ms) {
      try { Thread.sleep(ms); }
      catch (InterruptedException e) { }
  }
  public static void iAmAlive() {
      // blink LED on TINI board
    (new Thread() {
        BitPort bp=new BitPort(BitPort.Port3Bit5);
          public void run() {
```

Listing 4-8

Continued

```
            while (true) {
                bp.clear();
                pause(250);
                bp.set();
                pause(250);
            }
        }
    }).start();
}
synchronized static public boolean processRequest(char c,PrintWriter os) {
 if (c=='*') return false;
 if (c=='1') {
    stateRequest=true;
    cmdRequest=true;
 }
 if (c=='0') {
        stateRequest=false;
        cmdRequest=true;
 }
 if (c=='?') os.println(state?"1":"0");
 return true;
 }
 public static void main(String args[]) throws Exception {
     int i;
  System.out.println("Here we go!");
  iAmAlive();
    NetServer svr = new NetServer(111);
  svr.start();
  DSPortAdapter pa = OneWireAccessProvider.getDefaultAdapter();
  pa.targetFamily(0x12); // find DS2406
i=0;
   do {
     try {
        for (Enumeration e=pa.getAllDeviceContainers();e.hasMoreElements();i++) {
            buttons[i]=(OneWireContainer12) e.nextElement();
```

Listing 4-8
Continued

```
            }
        }
            catch (Exception e) { System.out.println("Exception: "+e);}
} while (i<4);
boolean b=true;
for (i=0;i<4;i++) {
ary[i]=buttons[i].readDevice();
 buttons[i].setLatchState(0,false,false,ary[i]);
buttons[i].writeDevice(ary[i]);
}
while (true)  {
 if (cmdRequest) {
      cmdRequest=false;
      setLED(2,stateRequest);
 }
ary[0]=buttons[0].readDevice();
 if (!buttons[0].getLevel(0,ary[0])) {
      setLED(3,true);
      state=true;
   System.out.println("On");
   }
 ary[1]=buttons[1].readDevice();
   if (!buttons[1].getLevel(0,ary[1])) {
         setLED(3,false);
         state=false;
   System.out.println("Off");
   }
 }
}
synchronized static void setLED(int n,boolean state) throws Exception {
   ary[n]=buttons[n].readDevice();
   buttons[n].setLatchState(0,state,true,ary[n]);
   buttons[n].writeDevice(ary[n]);
   // Inform outside world
   try {
```

Listing 4-8
Continued

```
    Socket sock = new Socket("192.168.0.1",112);
    String s = "Change: " + n + ":" + state + "\r\n";
    OutputStream os=sock.getOutputStream();
    os.write(s.getBytes());
    sock.close();
  }
  catch (Exception e) {
  }
 }
}
```

Listing 4-8
Continued

Switch2 is almost identical to Switch1, with the exception of the *setLED* method. This routine does all the work when either of the output LEDs changes state. Therefore, it is easy enough to add five lines of code to write a status message out to a socket.

The corresponding PC server appears in Listing 4-9. Since the message consists of a single line and presumably there is only one TINI, I didn't bother making this server spawn threads. Instead, each call to *accept* receives exclusive processing until it completes.

High-Level Notifications

One of the nice things about TINI is that it supports Java so well you can often develop code on the PC and then port it to the TINI with little or no effort. For example, consider the *SMTP* class from Chap. 3. This class uses *join*, which early versions of TINI did not support. However, if you are using the latest version, you'll be able to use this class with no problems.

It is a simple matter to replace the code that sends notifications to a socket with a call to the *SMTP* object. You can then send an e-mail instead. You can find the resulting code in Listing 4-10.

One of the problems with this code is that it is relatively large for the basic TINI. The TINI has 512K of memory, but you are sharing that

```java
import java.net.*;
import java.io.*;
public class Switch2Test {
 public static void main(String args[]) throws Exception {
     ServerSocket sock = new ServerSocket(112);
     while (true) {
       try {
             Socket skt = sock.accept();
             BufferedReader rdr = new BufferedReader(new InputStreamReader(skt.getInput
              Stream())));
             String s=rdr.readLine();
             System.out.println("Rcvd: " + s);
             skt.close();
       }
       catch (Exception e) { }
     }
   }
}
```

Listing 4-9
Reading Switch Messages on the PC

```java
import java.util.*;
import java.net.*;
import java.io.*;
import com.dalsemi.onewire.OneWireAccessProvider;
import com.dalsemi.onewire.adapter.DSPortAdapter;
import com.dalsemi.onewire.container.OneWireContainer12;
import com.dalsemi.system.BitPort;
class ProcessCmd extends Thread {
 Socket sock;
 public ProcessCmd(Socket sock) { this.sock=sock; }
 public void run() {
     try {
         BufferedReader is = new BufferedReader(new InputStreamReader(sock.getInputStream()));
         PrintWriter os=new PrintWriter(new OutputStreamWriter(sock.getOutputStream()),true);
```

Listing 4-10
E-mailing Status Messages.

```
        while (true) {
            String line=is.readLine();
            System.out.println("Read: "+line);
            if (!Switch3.processRequest(line.charAt(0),os)) {
                sock.close();
                return;
            }
        }
    } catch (Exception e) {
        try { if (sock!=null) sock.close(); }
        catch (IOException e0) {}
        System.out.println("Exception while processing cmd: " + e);
    }
  }
}
class NetServer extends Thread {
  static ServerSocket server=null;
  static int portno;
  NetServer(int port) { portno=port; }
  public void run() {
    if (server==null) {
        try {
            server=new ServerSocket(portno);
            while (true) {
              Socket newsock=server.accept();
              ProcessCmd cmd=new ProcessCmd(newsock);
              cmd.start();
            }
        } catch (Exception e) {
            System.out.println("Can't start server: " + e);
        }
    }
  }
}
public class Switch3 {
 static boolean state=false;
 static boolean cmdRequest = false;
```

Listing 4-10
Continued

```
static boolean stateRequest;
static OneWireContainer12[] buttons = new OneWireContainer12[4];
static byte[][] ary = new byte[4][];
public static void pause(int ms) {
    try { Thread.sleep(ms); }
    catch (InterruptedException e) { }
}
public static void iAmAlive() {
    // blink LED on TINI board
  (new Thread() {
        BitPort bp=new BitPort(BitPort.Port3Bit5);
            public void run() {
                while (true) {
                    bp.clear();
                    pause(250);
                    bp.set();
                    pause(250);
                }
            }
    }).start();
}
synchronized static public boolean processRequest(char c,PrintWriter os) {
if (c=='*') return false;
if (c=='1') {
   stateRequest=true;
   cmdRequest=true;
}
if (c=='0') {
    stateRequest=false;
    cmdRequest=true;
}
if (c=='?') os.println(state?"1":"0");
return true;
}
public static void main(String args[]) throws Exception {
    int i;
  System.out.println("Here we go!");
```

Listing 4-10
Continued

```
iAmAlive();
  NetServer svr = new NetServer(111);
svr.start();
   DSPortAdapter pa = OneWireAccessProvider.getDefaultAdapter();
   pa.targetFamily(0x12); // find DS2406
 i=0;
   do {
     try {
        for (Enumeration e=pa.getAllDeviceContainers();e.hasMoreElements();i++) {
              buttons[i]=(OneWireContainer12) e.nextElement();
           }
        }
              catch (Exception e) { System.out.println("Exception: "+e);}
   } while (i<4);
   boolean b=true;
   for (i=0;i<4;i++) {
   ary[i]=buttons[i].readDevice();
      buttons[i].setLatchState(0,false,false,ary[i]);
   buttons[i].writeDevice(ary[i]);
   }
   while (true)  {
     if (cmdRequest) {
         cmdRequest=false;
         setLED(2,stateRequest);
     }
   ary[0]=buttons[0].readDevice();
     if (!buttons[0].getLevel(0,ary[0])) {
         setLED(3,true);
         state=true;
     System.out.println("On");
     }
   ary[1]=buttons[1].readDevice();
     if (!buttons[1].getLevel(0,ary[1])) {
         setLED(3,false);
         state=false;
     System.out.println("Off");
```

Listing 4-10
Continued

```
        }
    }
}
synchronized static void setLED(int n,boolean state) throws Exception {
    ary[n]=buttons[n].readDevice();
    buttons[n].setLatchState(0,state,true,ary[n]);
    buttons[n].writeDevice(ary[n]);
    // Inform outside world
    System.out.println("Sending mail");
  String s = "Change: " + n + ":" + state + "\r\n";
    MailMessage msg=new MailMessage("TINI@al-williams.com","alw@al-williams.com","Test",s);
    System.out.println("Creating SMTP");
  SMTP smtp=new SMTP("bardo.clearlight.com");
    System.out.println("Status="+smtp.sendMail(msg));
  }
}
```

Listing 4-10
Continued

memory with slush, the file system, and some of the basic code that operates the TINI. If you've been doing a lot of development, you may get an out-of-memory exception when you try to execute this program, which is easily the largest program so far.

If you run out of memory, there are a few things you can do. First, delete unnecessary programs from the TINI's file system. Also, issue the slush command *gc* to force garbage collection. You can check the free memory condition by issuing a *df* command. You'll notice that running *gc* several times in a row usually frees up a little more memory each time.

The Other Side of E-mail

If the TINI can send e-mail to report its status, it makes sense that you might want to send TINI commands by e-mail. Normally an e-mail program will forward a message to an SMTP server, which will forward it—directly or indirectly—to the machine that handles mail for that domain. Normally, the receiving computer is a server that is on and connected to the Internet most of the time so that it can always respond to incoming mail requests.

What about computers like TINI that might have other things to do? A remote TINI might have to dial up to the Internet, so how can it accept mail messages? This is a common problem since dialup PCs are also in this category. The answer is to use a mail protocol like POP3 (defined in RFC1939). SMTP is a protocol for sending and transferring mail. POP3 is for delivering mail.

The idea is simple. When a user receives mail, the server computer accepts it on the user's behalf and then stores it. When the user is ready to check for incoming mail, he or she signs on to the POP3 server using a user ID and password. The user can then download the incoming messages and optionally delete them from the server. This basic method is how nearly all Internet users receive e-mail (although some use a newer protocol named IMAP, which is similar).

Can the TINI interact with a POP3 server? Of course. However, since e-mail may not immediately reach its destination, e-mail is not suitable for control in real time. On the other hand, a POP3 server will hold e-mail for the TINI indefinitely so that if the TINI is down, messages will be waiting when it returns to service.

Like other protocols, there are several ready-to-use classes for POP3. I especially liked the open source implementation I found at www.geocities.com/SunsetStrip/Studio/4994/java.html. These classes implement the POP3 protocol and are easy to use.

The *pop* class provides a handful of methods that you'll often use (see Table 4-1). Usually, you'll specify a server, user ID, and password in the constructor. This will determine which mail host the object uses. The calls typically return a *popStatus* object. Many of the methods also return multiple strings. You can examine these strings by calling the *popStatus* object's *Responses* method and walking through the array.

Listing 4-11 shows a simple class that scans a POP3 server for e-mail. Here's the basic steps it takes:

1. It logs in to the POP3 server.
2. It retrieves a list of e-mail messages.
3. It reads each message's headers searching for a specific e-mail address (tini@al-williams.com in this case).
4. If the e-mail message matches, the code calls *action*. The main *action* routine splits the message headers into a *hashtable* and collects the message body into a single string. Then it calls another version of

TABLE 4-1	Method	Description
Common Methods for the *pop* Class	connect	Connects to the specified server.
	dele	Marks a numbered message for deletion.
	list	Obtains a list of valid message numbers and their sizes.
	login	Logs in to the server.
	quit	Ends the session and deletes any messages marked for deletion.
	retr	Retrieves a message.
	rset	Cancels message deletion.

```java
import com.jthomas.pop.*;
import java.util.Hashtable;
public class poplist {
 static final String address="tini@al-williams.com";
 public static String getAddress() { return address; }
 public static void main(String arg[]) {
   pop3 pop = new pop3(arg[0], arg[1], arg[2]);
   popStatus status = pop.connect(), xstatus;
   if ( status.OK() )
     status = pop.login();
   if ( status.OK() ) {
     status = pop.list();
     String[] responses = status.Responses();
       String[] xresponses;
       int n;
     for(int i=0; i< responses.length; i++) {
       n=responses[i].indexOf(' ');
       n=Integer.parseInt(responses[i].substring(0,n));
       xstatus=pop.retr(n);
       if (xstatus.OK()) {
         xresponses = xstatus.Responses();
```

Listing 4-11
Simple Class That Scans a POP3 Server for E-Mail

```
            for (int n1=0;n<xresponses.length;n1++) {
                if (xresponses[n1].length()==0 ) break;
                if (!xresponses[n1].substring(0,3).equalsIgnoreCase("to:")) continue;
                if (xresponses[n1].indexOf(getAddress())==-1) continue;
                // this e-mail message is for me
                // do something with e-mail message #m
                // including delete it most likely
                action(pop,n);
                pop.dele(n); // kill message
            }
        } else { System.out.println("RETR ERROR"); }
    }
    status = pop.quit(); // must quit to delete properly
  }
}
// This is the low-level action that, by default,
// builds a hashtable of the headers and a string of the message
// only override this if you want greater control
public static void action(pop3 pop, int n) {
    boolean header=true;
    popStatus status=pop.retr(n);
    if (!status.OK()) return;
    String [] responses = status.Responses();
    StringBuffer message = new StringBuffer();
    Hashtable headers = new Hashtable();
    for (int i=0;i<responses.length;i++) {
        if (responses[i].length()==0) {
            header=false;
            continue;
        }
        if (header) {
            int n1=responses[i].indexOf(':');
            if (n1!=-1) headers.put(responses[i].substring(0,n1).trim(),responses[i].
              substring(n1+1).trim());
        }
```

Listing 4-11
Continued

```
        else {
              message.append(responses[i]);
              message.append("\n");
        }
    }
    action(headers,message.toString());
}
// This is the action routine you will generally override
public static void action(Hashtable headers, String message) {
    System.out.println("Subject is " + (String)headers.get("Subject"));
    System.out.println("Headers\n" + headers);
    System.out.println("----");
    System.out.println(message);
}
}
```

Listing 4-11
Continued

action. You can change this version of the method and perform whatever processing you like.

The raw e-mail message format contains headers, a blank line, and the message body. The headers all have a field name (like Subject), a colon, and the value of the field. The code examines the To field to determine the destination of the e-mail message. Also, the program only deletes messages that it recognizes. This is useful since many POP3 servers will collect all mail for a single domain (like al-williams.com). With this scheme, multiple programs can share a POP3 server as long as they don't collect mail that isn't specifically addressed to them. For example, a single POP3 server could provide mail for tini@al-williams.com, tini1@al-williams.com, and mastertini@al-williams.com. Each TINI would only read its own messages.

Goodbye Switch

The Switch program started out simply enough. It was a bare-bones flip-flop in software. After a few revisions—and the addition of a few pieces of related workstation software—it can accept commands via the Internet, notify other programs, or even send you an e-mail. While this may seem simplistic, there are many practical applications. Would you like to get an e-mail (or even an alphanumeric page via e-mail) when your basement gets wet? Would you like an instant message if the front door to your office opens? The programs needed for these projects would not be much different from the basic Switch program.

Of course, since I wanted to emphasize the changes in the program, the final program here is a bit bare-bones. In real life, you'd want to make the program a bit more configurable. This is relatively simple to do with Java properties.

Ideally, the configuration data would reside in a database. With the TINI you need to keep things simple, but a Java property file will fit the bill. For this purpose, it is all the database you need.

One advantage to property files is that they're easy to create with an ordinary text editor. It is straightforward to access them from a servlet (something you'll read about in Chap. 5).

To access a property file, you create an *InputStream* object (using, for example, *FileInputStream*) and pass it to a *Properties* object's load method. Keep in mind that within a JSP file you want to open the file relative to the server's root. The trick is to use the *getServletContext* method to retrieve a *ServletContext* object. You can then call the *getResourceAsStream* method to convert a server-relative URL into an input stream.

Listing 4-12 shows a typical property file. The hash (#) character indicates a comment, so the text following it is not really part of the database. Each name and string pair represents an entry in the property database.

Listing 4-12
Typical Property File

```
serverPort=111
debugMsg=true
serverNotify=false
serverNotifyHost=192.168.0.1
serverNotifyPort=112
mailNotify=true
smtpServer=smtp.yahoo.com
from=TINI1@al-williams.com
to=alw@al-williams.com
subj=Data from TINI
```

To read a property object from a file, you'll create an *InputStream* object (using, for example, *FileInputStream*) and pass it to a *Properties* object's *load* method. Listing 4-13 is a very simple program that reads a property file and dumps a value from it.

You can find a complete sample of using property files with the *Switch* code in Listing 4-14. Via the property file you can change the host names and port numbers, and also select between e-mail notification, socket-based notification, or both.

Listing 4-13
Simple Program to
Read a Property
Object from a File

```java
import java.util.*;
import java.io.*;
public class PropTest {
  public static void main(String[] args) throws Exception {
      FileInputStream fin=new FileInputStream("test.properties");
      Properties prop = new Properties();
      prop.load(fin);
      System.out.println(prop.getProperty("test"));
  }
}
```

```java
import java.util.*;
import java.net.*;
import java.io.*;
import com.dalsemi.onewire.OneWireAccessProvider;
import com.dalsemi.onewire.adapter.DSPortAdapter;
import com.dalsemi.onewire.container.OneWireContainer12;
import com.dalsemi.system.BitPort;
class ProcessCmd extends Thread {
  Socket sock;
  public ProcessCmd(Socket sock) { this.sock=sock; }
  public void run() {
    try {
        BufferedReader is = new BufferedReader(new InputStreamReader(sock.getInputStream()));
```

Listing 4-14
Setting Options with Property Files

```
        PrintWriter os=new PrintWriter(new
         OutputStreamWriter(sock.getOutputStream()),true);
        while (true) {
            String line=is.readLine();
            System.out.println("Read: "+line);
            if (!Switch4.processRequest(line.charAt(0),os)) {
               sock.close();
               return;
            }
        }
     } catch (Exception e) {
        try { if (sock!=null) sock.close(); }
        catch (IOException e0) {}
        System.out.println("Exception while processing cmd: " + e);
     }
   }
}
class NetServer extends Thread {
  static ServerSocket server=null;
  static int portno;
  NetServer(int port) { portno=port; }
  public void run() {
     if (server==null) {
        try {
              server=new ServerSocket(portno);
              while (true) {
                Socket newsock=server.accept();
                ProcessCmd cmd=new ProcessCmd(newsock);
                cmd.start();
              }
         } catch (Exception e) {
              System.out.println("Can't start server: " + e);
         }
     }
   }
```

Listing 4-14
Continued

```
}
public class Switch4 {
 static boolean state=false;
 static boolean cmdRequest = false;
 static boolean stateRequest;
 static OneWireContainer12[] buttons = new OneWireContainer12[4];
 static byte[][] ary = new byte[4][];
 // Property parameters
 static String props="Switch4.properties";
 static int serverPort=111;
 static boolean debugMsg=false;
 static boolean serverNotify=false;
 static String serverNotifyHost="192.168.0.1";
 static int serverNotifyPort = 112;
 static boolean mailNotify=true;
 static String smtpServer="bardo.clearlight.com";
 static String from="TINI@al-williams.com";
 static String to="alw@al-williams.com";
 static String subj="Test";
 // Read string, boolean, or integer properties with defaults
 private static String getAProperty(Properties p,String key,String deflt) {
    String rv=p.getProperty(key);
       if (rv==null || rv.equals("")) rv=deflt;
       return rv;
 }
 private static boolean getAProperty(Properties p,String key,boolean deflt) {
    String rv=p.getProperty(key);
       if (rv==null || rv.equals("")) return deflt;
       // some versions of TINI had incorrect getBoolean definition
       //    return Boolean.getBoolean(rv);
       return Boolean.valueOf(rv).booleanValue();
 }
 private static int getAProperty(Properties p,String key,int deflt) {
   String rv=p.getProperty(key);
      if (rv==null || rv.equals("")) return deflt;
```

Listing 4-14

Continued

```
            return Integer.parseInt(rv);
    }
    private static void readProperties() throws Exception {
        FileInputStream strm=new FileInputStream(props);
        Properties properties = new Properties();
        String tmp;
        properties.load(strm);
        serverPort=getAProperty(properties,"serverPort",serverPort);
        debugMsg=getAProperty(properties,"debugMsg",debugMsg);
        serverNotify=getAProperty(properties,"serverNotify",serverNotify);
        serverNotifyHost=getAProperty(properties,"serverNotifyHost",serverNotifyHost);
        serverNotifyPort=getAProperty(properties,"serverNotifyPort",serverNotifyPort);
        mailNotify=getAProperty(properties,"mailNotify",mailNotify);
        smtpServer=getAProperty(properties,"smtpServer",smtpServer);
        from=getAProperty(properties,"from",from);
        to=getAProperty(properties,"to",to);
        subj=getAProperty(properties,"subj",subj);
        if (debugMsg) System.out.println("Properties loaded");
    }
    public static void pause(int ms) {
        try { Thread.sleep(ms); }
        catch (InterruptedException e) { }
    }
    public static void iAmAlive() {
        // blink LED on TINI board
      (new Thread() {
            BitPort bp=new BitPort(BitPort.Port3Bit5);
              public void run() {
                while (true) {
                    bp.clear();
                    pause(250);
                    bp.set();
                    pause(250);
                }
            }
```

Listing 4-14
Continued

```
        }).start();
  }
  synchronized static public boolean processRequest(char c,PrintWriter os) {
   if (c=='*') return false;
   if (c=='1') {
      stateRequest=true;
      cmdRequest=true;
   }
   if (c=='0') {
        stateRequest=false;
        cmdRequest=true;
   }
   if (c=='?') os.println(state?"1":"0");
   return true;
  }
  public static void main(String args[]) throws Exception {
      int i;
      readProperties();
    if (debugMsg) System.out.println("Here we go!");
    iAmAlive();
      NetServer svr = new NetServer(serverPort);
    svr.start();
      DSPortAdapter pa = OneWireAccessProvider.getDefaultAdapter();
      pa.targetFamily(0x12); // find DS2406
    i=0;
      do {
        try {
          for (Enumeration e=pa.getAllDeviceContainers();e.hasMoreElements();i++) {
              buttons[i]=(OneWireContainer12) e.nextElement();
          }
        }
              catch (Exception e) { System.out.println("Exception: "+e);}
      } while (i<4);
      boolean b=true;
      for (i=0;i<4;i++) {
```

Listing 4-14
Continued

```
    ary[i]=buttons[i].readDevice();
      buttons[i].setLatchState(0,false,false,ary[i]);
    buttons[i].writeDevice(ary[i]);
    }
    while (true)  {
      if (cmdRequest) {
            cmdRequest=false;
            setLED(2,stateRequest);
      }
    ary[0]=buttons[0].readDevice();
        if (!buttons[0].getLevel(0,ary[0])) {
            setLED(3,true);
            state=true;
        if (debugMsg) System.out.println("On");
        }
      ary[1]=buttons[1].readDevice();
        if (!buttons[1].getLevel(0,ary[1])) {
            setLED(3,false);
            state=false;
        if (debugMsg) System.out.println("Off");
      }
  }
}
    synchronized static void setLED(int n,boolean state) throws Exception {
    ary[n]=buttons[n].readDevice();
    buttons[n].setLatchState(0,state,true,ary[n]);
    buttons[n].writeDevice(ary[n]);
    // Inform outside world
      String s = "Change: " + n + ":" + state + "\r\n";
    if (mailNotify) {
      if (debugMsg) System.out.println("Sending mail");
      MailMessage msg=new MailMessage(from,to,subj,s);
      SMTP smtp=new SMTP(smtpServer);
      if (debugMsg) System.out.println("Status="+smtp.sendMail(msg));
      }
```

Listing 4-14

Continued

```
if (serverNotify) {
    try {
        Socket sock=new Socket(serverNotifyHost,serverNotifyPort);
        OutputStream os=sock.getOutputStream();
        os.write(s.getBytes());
        sock.close();
    }
    catch (Exception e) {
    }
  }
 }
}
```

Listing 4-14
Continued

To make working with properties easier, the program defines a few helper functions. First, *readProperties* encapsulates all the property-related code so that there isn't a big impact on the existing code. The program also defines three versions of *getAProperty*. These functions are practically identical, except they each handle a different data type (*String, boolean,* or *int*). You pass the function the *Properties* object, the key, and a default value of the appropriate type. This default value is how Java differentiates between the three functions. You can't define overloaded functions that only differ in their return type.

Now it is simple to deploy the Switch program and customize it for the particular environment without having to recompile each time. You could even make the program accept a property file name on its command line to allow running multiple copies on each TINI.

When Java's Not Enough

On the face of it, TINI seems ideal. It is small, inexpensive, and runs Java, which makes it easy to program. Why use anything else? The problem with TINI is that, while it is small and inexpensive, it might not be small enough or inexpensive enough. If you are wiring up three or four data loggers, the TINI is perfect. However, if you are trying to build things in volume, or to build very small devices, the TINI may still be too expensive and too large.

At times like this, you may have to turn to an even smaller microcontroller. Of course, a smaller microcontroller will need more support hardware and will probably be harder to program. In addition, you may need a special programmer and possibly some development tools. In Chap. 6, you'll see an easy way to use most of what you know about Java to achieve low-level control.

All for TINI?

Now that you have finished this chapter, you should be comfortable with the basic operation of the TINI. The programs have ranged from a blinking light to various socket-based transactions to and from the remote network computer.

However, there is one thing still unexplored: the Web. An e-mail notifying you of an event is nice, but what about viewing that status on a Web page served by the TINI? That's what you'll learn how to do in Chap. 5.

CHAPTER 5

TINI Meets the Web

How hard is it to make a phone call between, say, the United States and Malaysia? Not hard at all. Now, how hard is it to have a meaningful conversation with someone in that country? That depends. If you can find an American that speaks Malay (the national language of Malaysia) or a Malaysian that speaks English, that's great. However, if the two parties don't share a common language, the call itself is of little value.

In Chap. 4 you found that the TINI could easily open up a network connection or wait for others to connect to it. This is—fundamentally—what Web browsers and servers do. Of course, it isn't enough to make the connection; you also have to speak the language.

There are at least three strategies you can use when you want to connect TINI to the Web. These are generally similar to the options you'll probably have on any similar Java-based processor.

The first option available is to use the TINI's built-in Web server class to create a customized server. The problem with this is that it isn't always easy to extend this class to do exactly what you want since all the work is done outside of your program.

Another possibility is to create your own Web server from scratch. That may sound crazy, but it turns out that Java is very well suited to writing a Web server and it isn't nearly the chore you'd imagine. What's more, there are plenty of open-source Web servers that you can use as a basis for your customized versions.

The final alternative is a variation of the second choice. There is a special open-source Web server available for the TINI that also supports Java servlets. Servlets are small Java programs that the Web server can dynamically load and integrate. That means you can customize this Web server on the fly very easily. Many full-blown Web servers support servlets, but it is somewhat surprising that a little computer like TINI can provide them as well.

Which method is best? That depends, as always, on what you are trying to do. However, you'll get a look at all three techniques in this chapter.

The Built-In Solution

TINI's default library has a class named *HTTPServer*. This class isn't very useful by itself, but with just a bit of coaxing you can turn it into an actual Web server. The only problem with this class is that it does so much for you it is hard to change its operation. Once you set it up, you simply call *serviceRequests* and it does all the work. That's great if you want it to work exactly as it was designed.

The problem is that most TINI Web servers won't serve static pages. After all, if you just want a file on the Web, there are thousands of host computers you could store it on—why use a TINI? The strength of the TINI is that it can monitor and control real-world events. Unfortunately, those events don't typically reside in an HTML file.

In the TINI example files, there is an example server that reports the time and temperature (assuming you have a temperature sensor attached). It runs a separate thread that periodically updates an external file that the server then provides to browsers. While this is sometimes acceptable, you'll more often want to alter the contents of a Web page on the fly, not periodically.

This also leads to a problem where the Web server thread can't access the file while the updating thread has it open. To prevent problems, you must synchronize the two threads so they will not access files at the same time.

An Example Server

Consider Listing 5-1. It shows a class that extends *HTTPServer*. It also implements *Runnable* since it will use a thread to satisfy Web requests.

```
/*
*AlWebWorker.java
*/
import java.io.*;
import com.dalsemi.tininet.http.*;
//import com.dalsemi.nethack.*;
/** This class runs a web server.
 */
public class AlWebWorker extends HTTPServer
  implements Runnable
{
```

Listing 5-1
Class That Extends *HTTP Server* and Implements *Runnable*

```
Object  lock; // lock for file access
String  threadName;
byte[]  name;
  // HTTPServer    httpServer;
public int     httpPort   = HTTPServer.DEFAULT_HTTP_PORT;
public String webRoot     = "/";
public String webIndex    = "index.html";
boolean        debugOn    = false;
/** Constructor
  */
public AlWebWorker(Object lock)
{
  try
  {
    this.lock = lock;
    // override the default index page
    setIndexPage(webIndex);
    // override the default HTTP root
    setHTTPRoot(webRoot);
  }
  catch(HTTPServerException h)
  {
    if(debugOn)
    {
      System.out.println(h.toString());
    }
  }
  boolean  loggingFailed = false;
  try
  {
    // could enable logging here if you wanted to
    setLogging(false);
  }
  catch(HTTPServerException h)
  {
```

Listing 5-1
Continued

```
      // problem with log file
      loggingFailed = true;
      if(debugOn)
      {
        System.out.println(h.toString());
      }
    }
    try
    {
      if(loggingFailed)
        setLogging(false);
    }
    catch(HTTPServerException h)
    {
      if(debugOn)
      {
        // no need to do anything here? if we can't close it, we can't close it
        System.out.println(h.toString());
      }
    }
  }
  /** Run the server thread
    */
  public void run()
  {
    threadName = Thread.currentThread().getName();
    name       = (threadName+"\n").getBytes();
    if(debugOn)
    {
      System.out.println(threadName);
    }
    int result = 0;
    while(true)
    {
      try
```

Listing 5-1

Continued

```
    {
        // Threaded web server blocks on accept
      result = serviceRequests(lock);
      if(debugOn)
      {
        System.out.println(new String("<"+result+">"));
      }
    }
    catch(HTTPServerException h)
    {
      if(debugOn)
      {
        System.out.println(h.toString());
      }
    }
    catch(Throwable t)
    {
      if(debugOn)
      {
        // why kill the server if the exception is not fatal?
        System.out.println(t.toString());
      }
    }
  }
 }
}
```

Listing 5-1
Continued

The constructor requires an object that will supply the monitor to prevent the server from accessing files while the update thread is executing. The constructor also overrides the default Web directory (that is, the directory where the server will find Web pages) and the default document.

The *run* function is, of course, the threaded part of the object. This does little more than call the underlying *serviceRequests* function, which takes care of all the details.

Listing 5-2 shows a program that uses this Web server object. This program accepts command line arguments, if any, and creates the Web server thread with the appropriate options.

When the server starts, it calls *createPage*. This method looks for a local file named indextop.html. If it exists, then control passes to *createLocalPage*. This method joins the contents of indextop.html, the return

```
/*
 * AlWebServer.java
 */
import java.net.*;
import java.io.*;
import java.util.*;
import com.dalsemi.system.*;
public class AlWebServer
{
 AlWebWorker        webWorker;
 byte[]  copyBuffer;
 String  webRoot,
         webIndex;
 boolean localPages       = false;
 static  Object lock      = new Object();// lock for file access
 Clock   clock            = new Clock();
 static String indexTop   = "<HTML>"+
                "<HEAD>"+
                "<TITLE>Al's WebServer</TITLE>"+
                "</HEAD>"+
                "<BODY COLOR=\"#0000FF\" BGCOLOR=\"#FFFFFF\" ALINK=\"#C0C0C0\">"+
                "<META HTTP-EQUIV=\"Expires\" CONTENT=\"0\">"+
                "<META HTTP-EQUIV=\"Last modified\" CONTENT=\"now\">"+
                "<META HTTP-EQUIV=\"Pragma\" CONTENT=\"no-cache\">"+
                "<META HTTP-EQUIV=\"Cache-Control\" CONTENT=\"no-cache, must-vali
                 date\">"+
                "<FONT ALIGN=\"CENTER\" COLOR=\"#0000FF\">"+
                "<H1 ALIGN=\"CENTER\">AlWebServer</H1>"+
                "<H3 ALIGN=\"CENTER\">If you can read this, Al's WebServer is
```

Listing 5-2
Program That Creates Web Server Thread

```
running<BR>On TINI</H3><BR>"+
                    "<P ALIGN=\"CENTER\">"+
                    "The AlWebServer application uses the HTTPServer class to implement a
simple web server.<BR>"+
    "</P>";
  static String indexBottom  =        "</H1><BR>"+
                    "</FONT>"+
                    "</BODY>"+
                    "</HTML>";
  /**  Constructor
   */
  public AlWebServer()
   throws IOException
 {
   copyBuffer = new byte[1024];
}
   // Center portion of index file
 String getContent() {
     clock.getRTC();
     int hour  = clock.getHour();
     int minute  = clock.getMinute();
     int second  = clock.getSecond();
     String timeString = new String("<BR>Current time " + hour + ":");
   if(minute < 10)
     timeString += "0";
   timeString += Integer.toString(minute);
   timeString += ":";
   if(second < 10)
     timeString += "0";
   timeString += Integer.toString(second);
   return timeString;
   }
 public void createPageLocal()
 {
  try
  {
```

Listing 5-2
Continued

```
     String content=getContent();
     synchronized(lock)
     {
       FileOutputStream  index = new FileOutputStream(new File(webRoot, webIndex));
       index.write(indexTop.getBytes(), 0, indexTop.length());
       index.write(content.getBytes());
       index.write(indexBottom.getBytes(), 0, indexBottom.length());
       index.close();
     }
   }
   catch(Exception e)
   {
     System.out.println("createPageLocal -"+e.toString());
   }
}
public void createPage()
{
 try
   {
    File indexPage = new File(webRoot+"indextop.html");
    if(!indexPage.exists())
    {
     localPages = true;
     createPageLocal();
       return;
      }
     String content = getContent();
     synchronized(lock)
     {
        FileOutputStream  index      = new FileOutputStream(new File(webRoot, webIndex));
        FileInputStream   tempFile   = null;
        int bytesRead = (tempFile = new FileInputStream(new
File(webRoot,"indextop.html"))).read(copyBuffer);
        index.write(copyBuffer, 0, bytesRead);
        tempFile.close();
        index.write(content.getBytes());
```

Listing 5-2
Continued

```
      bytesRead = (tempFile = new FileInputStream(new
File(webRoot,"indexbottom.html"))).read(copyBuffer);

      index.write(copyBuffer, 0, bytesRead);

      tempFile.close();

      index.close();

    }
  }
  catch(Exception e)
  {
    System.out.println("createPage -"+e.toString());
  }
}
/** Start the various servers. Then check and
 * update the web page.
 */
public void startup()
{
  try
  {
    // create the web server
    Thread webServer = new Thread(webWorker);
    webServer.setName("Web Server");
    webRoot   = webWorker.getHTTPRoot();
    webIndex  = webWorker.getIndexPage();
    createPage();
    // start the web server
    webServer.start();
    // periodically wake up and refresh page
    while(true)
    {
      Thread.sleep(5000);
        createPage();
    }
  }
  catch(Throwable t)
  {
```

Listing 5-2

Continued

```
      System.out.println("createPageLocal -"+t.toString());
      // why kill the server if the exception is not fatal?
      //System.out.println(t);
   }
}
/** Print a usage statement
  */
void printUsage()
{
   System.out.println("Usage: AlWebServer [-p portNum] [-i indexPage] [-r webRoot] [-d]");
   System.out.println("      p use specified port");
   System.out.println("      i use specified index page");
   System.out.println("      r use specified directory as web root");
}
/** Create a AlWebServer.
  */
public static void main(String[] args)
{
   System.out.println("Starting Al WebServer version 2.0 ...");
   System.out.println("Based on TINIWebServer example");
   try
   {
      int currentArg    = 0;
      AlWebServer alWebServer    = new AlWebServer();
      alWebServer.webWorker        = new AlWebWorker(alWebServer.lock);
      if((args != null)&&(args.length > 0))
      {
         while((currentArg < args.length) &&
            (args[currentArg] != null) &&
            (args[currentArg].length() != 0))
         {
            if(args[currentArg].charAt(0) != '-')
            {
               alWebServer.printUsage();
               return;
```

Listing 5-2

Continued

```
      }
      switch(args[currentArg++].charAt(1))
      {
        case 'p':
          int portDesired = Integer.valueOf(args[currentArg++]).intValue();
          if(portDesired > 0)
            alWebServer.webWorker.httpPort = portDesired;
        break;
        case 'i':
          alWebServer.webWorker.webIndex = args[currentArg++];
        break;
        case 'r':
          alWebServer.webWorker.webRoot = args[currentArg++];
        break;
        default:
          alWebServer.printUsage();
        return;
      }
    }
  }
  alWebServer.startup();
  }
  catch(Throwable t)
  {
    System.out.println(t);
    System.out.println(t.getMessage());
  }
  finally
  {
    System.out.println("\nAlWebserver exiting...");
  }
 }
}
```

Listing 5-2
Continued

value of *getContent,* and indexbottom.html to form a new index.html file. If the files do not exist, the *createPage* method uses some default text for the top and bottom (contained in the *indexTop* and *indexBottom* variables). In either case, the middle portion of the file is the return value of *getContent.*

Once the server starts, it enters a loop where it sleeps for 5 s and then calls *createPage* again. Each call to *createPage* acquires the lock that the program shares with the Web server. That means that if the server is busy, *createPage* will wait and vice versa. The net effect is that the index.html page changes every 5 s (approximately) to reflect the current time (or, at least, the current time according to TINI).

Of course, you could alter *getContent* to return any string you want, and even insert HTML into it. For example, you might get the status of switches (like the programs in Chap. 4) and display them as text or even graphics. Perhaps a picture of a red LED indicates off and a green LED indicates on. Just remember, the page only updates every 5 s on the server, so changes are not instantaneous. Also, the browser will not automatically reload the page just because it changes on the server. You'll have to reload the page if you want to see the changes.

You could automate the reloading on the browser by including a special tag in the header part of the HTML document:

```
<HEAD>
<META HTTP-EQUIV=REFRESH CONTENT=10>
```

In this case, the page will reload every 10 s. You would not want to reload every 5 s, because the server might not be finished updating the page yet. In general, you'll want to refresh the browser half as often as you update the Web page. Of course, you could safely update the browser less often as well.

What's Wrong with That?

The two files shown in Listings 5-1 and 5-2 form a complete Web server solution, and you didn't even have to know very much about writing a Web server to create it. Perfect, right? For some situations, it is a great way to serve a single dynamic page that doesn't need frequent updates.

However, what if you were trying to show multiple pages with data that should be up to date at all times? Updating 30 or 40 pages (and locking the

Web server in the process) would not be a good idea. Besides, to keep everything up to date, you'd have to update the files constantly.

A better idea would be to put some sort of placeholder in the HTML files. When the TINI server noticed the placeholder, it could call a method to substitute a dynamic value for that placeholder.

This sounds simple, but unfortunately, the *HTTPServer* class is not set up to handle this. However, it is not too difficult to write a full-blown server in Java, and then you can do what you want.

Handcrafted

Writing a Web server in Java is considerably easier than in most other languages. Still, there are some subtleties that you'd have to spend a little time resolving. Luckily, there are several Java-based open-source Web servers that can get you up and running in no time.

To keep things simple, I drew up a few ground rules for the server. First, it didn't need to support scripting, CGI, or the *POST* method, just the good old-fashioned HTTP 1.0 *GET* method. Second, I wanted the server to recognize requests for CGI scripts and do the processing in Java instead. Obviously, if you can code something so that it has general-purpose functions, that's even better.

My prototype pseudo-CGI script simply looks for a form entry named *Filename* (case is important). If the script finds this entry, it assumes that it should write out all the other entries in a comma-delimited format, appending it to the file named in *Filename*. Usually, *Filename* will be a hidden field on a form and set to some appropriate value by JavaScript running on the browser. Of course, you could override this with your own code. I also wanted to add a way for the server to replace placeholders as discussed earlier.

I looked at several Java-based Web servers, and I decided I liked one called *Webster* (http://richard5.net/projects/webster.php3). It has enough features to be worthwhile, but not so many that you can't follow the code (or stuff it into a TINI).

Webster didn't do everything I wanted it to do, but it was a good start. Besides, the author lets you copy it under the GNU public license, so you can reuse it with no trouble.

The server's code is remarkably short—fewer than 150 lines (even fewer if you take out blank lines and comments). However, the server doesn't correctly handle universal resource locator (URL) encoding (the process that browsers use to transform certain characters into their hex equivalents). This is not hard to fix (and may even be fixed by the time you read this).

Configuring the server depends on two files: server.properties and mimetypes.properties. The server.properties file sets the root directory for the server, the port number, and the default server file. The mimetypes.properties file sets the relationship between file extensions and document types (for example, .txt is text/plain and .htm is text/html).

To press the Webster object into service, I wanted to add three things:

1. Correct handling of URL encoded strings

2. Parsing of the query string passed to the server (for example, as part of a *FORM* with a *GET* action)

3. A function you could override to perform work after parsing the filename and query string, but before sending the file to the client

The first change was easy enough. While the *URL* object has a method for encoding strings, it doesn't do the reverse decoding. It's simple enough, however, to scan through the string replacing plus signs with spaces, and hex sequences (that is, hex numbers preceded by percent signs) with the correct equivalent. You can find the *URLDecode* method (with the rest of the server code) in Listing 5-3.

```
// Small HTTP Server by Al Williams
// This code is based on the Webster server at
// http://richard5.net/projects/webster.php3
// which is covered by the GNU General Public License
// detailed at http://www.gnu.org/copyleft/gpl.html
import java.net.*;
import java.io.*;
import java.util.*;
import com.dalsemi.system.Clock; // just for time
public class Server1 implements Runnable {
    private ServerSocket ss;
    private Thread runner=null;
    // The server's configuration information is stored in these properties
```

Listing 5-3
Web Server Class

```
    protected static Properties props = new Properties();
    // The mime types information is stored in this properties list
    protected static Properties MimeTypes = new Properties();
    Server1() {  // main constructor
      runner = new Thread(this);
      runner.start();
    }
    public static void main(String[] args) {
      new Server1();
    }
    // override this to provide an action
    public String action(String filename, Hashtable vars) {
      System.out.println("Serving: "+filename + " " + vars);
      return filename;
    }
  public void run() {
    try{
      loadProps();
      loadMimes();
      System.out.println("HttpServer listening on port:" +
              props.getProperty("portnumber"));
      // setup serversocket
      ss = new ServerSocket( (new Integer(props.getProperty
                ("portnumber"))).intValue() );
      while(true) {
        Socket s = ss.accept(); // accept incoming requests
        new Thread(new SendFile(this,s)).start();
      }
    } catch(Exception e) {
      System.out.println("Main Serve thread " + e );
    }
  }
  // load the properties file
  static void loadProps() throws IOException {
    File f = new File("server.properties");
```

Listing 5-3
Continued

```
     if (f.exists()) {
       InputStream is =
          new BufferedInputStream(new FileInputStream(f));
       props.load(is);
       is.close();
     }
 } // end of loadProps
 // load the properties file
 static void loadMimes() throws IOException {
    File f = new File("mimetypes.properties");
    if (f.exists()) {
       InputStream is =
          new BufferedInputStream(new FileInputStream(f));
       MimeTypes.load(is);
       is.close();
     }
   } // end of loadMimes
} // end of Serve class
class SendFile implements Runnable{
  private Socket client;
  private String fileName,header;
  private String query;
  private DataInputStream requestedFile;
  private int fileLength;
  private Server1 svr;
  SendFile(Server1 svr,Socket s) {  // constructor
    client = s;
    this.svr=svr;
  }
  String getTime() {
    Clock clock = new Clock();
      clock.getRTC();
      int hour   = clock.getHour();
      int minute = clock.getMinute();
      int second = clock.getSecond();
```

Listing 5-3
Continued

```java
      String timeString = new String("hour + ":");
    if(minute < 10)
      timeString += "0";
    timeString += Integer.toString(minute);
    timeString += ":";
    if(second < 10)
      timeString += "0";
    timeString += Integer.toString(second);
    return timeString;
  }
 void tiniSubstitute(int code,DataOutputStream str) throws IOException {
    if (code=='T') str.writeChars(getTime());
    else {
   str.writeByte('`'); // huh?
   str.writeByte(code);
  }
}
   public void run() {
     String line;
     try {
       BufferedReader dis =
           new BufferedReader(new InputStreamReader
                       (client.getInputStream()));
       // read request from browser and parse
       while((line=dis.readLine())!=null) {
         StringTokenizer tokenizer = new StringTokenizer(line," ");
         if (!tokenizer.hasMoreTokens()) break;
         if (tokenizer.nextToken().equals("GET")) {
           fileName = tokenizer.nextToken();
           if (fileName.endsWith("/")) {
             fileName = fileName +
                 Server1.props.getProperty("defaultfile");
           } else {
             fileName = fileName.substring(1);
           }
```

Listing 5-3

Continued

```
      }
   }
   if (fileName.charAt(0)!='/') fileName = "/" + fileName;
   int n=fileName.indexOf('?');
   if (n!=-1) {
      query=fileName.substring(n+1);
      fileName=fileName.substring(0,n);
   }
   else
      query="";
   fileName=URLDecode(fileName);
   // decode query string
   Hashtable qvars = new Hashtable(64);
   int n0,n1;
   do {
      String val,key;
      n0=query.indexOf('&');
      if (n0==-1) n0=query.length();
      if (n0<=0) break;
      String vpart = query.substring(0,n0);
      if (n0==query.length()) query="";
      else query=query.substring(n0+1);
      n1=vpart.indexOf('=');
      if (n1==-1) {
         val="";
         key=vpart;
      } else {
         val = vpart.substring(n1+1);
         key = vpart.substring(0,n1);
      }
      qvars.put(URLDecode(key),URLDecode(val));
   } while (!query.equals(""));
   fileName=svr.action(fileName,qvars);
   try {
      requestedFile = new DataInputStream(
```

Listing 5-3
Continued

```
        new BufferedInputStream (new
            FileInputStream(Server1.props.
            getProperty("root") + fileName)));
    fileLength = requestedFile.available();
    constructHeader();
} catch(IOException e) { // file not found send 404.
    header = "HTTP/1.0 404 File not found\n" +
        "Allow: GET\n" +
        "MIME-Version: 1.0\n"+
        "Server : HttpServer: a Java Local HTTP Server\n"+
        "\n\n <H1>404 File not Found</H1>\n";
    fileName = null;
}
int i;
DataOutputStream clientStream =
    new DataOutputStream(new BufferedOutputStream(client.
            getOutputStream()));
clientStream.writeBytes(header);
if (fileName != null) {
    while((i = requestedFile.read()) != -1) {
            if (i=='`') {
                    if ((i=requestedFile.read())==-1)break;
                    if (i!='`') {
                        tiniSubstitute(i,clientStream);
                        continue;
                    }
            }
        clientStream.writeByte(i);
    }
}
clientStream.flush();
clientStream.close();
dis.close();
client.close();
if (requestedFile!=null) requestedFile.close();
} catch(Exception e) {
```

Listing 5-3
Continued

```
        System.out.print("Error closing Socket\n"+e);
    }
}
public String URLDecode(String in)
{
  StringBuffer out = new StringBuffer(in.length());
  int i = 0;
  int j = 0;
  while (i < in.length())
  {
    char ch = in.charAt(i);   i++;
    if (ch == '+') ch = ' ';
    else if (ch == '%')
    {
       ch = (char) Integer.parseInt(
                in.substring(i,i+2), 16);
      i+=2;
    }
    out.append(ch);
    j++;
  }
  return new String(out);
}
private void constructHeader() {
   String fileType;
   fileType = fileName.substring(fileName.
          lastIndexOf(".")+1,fileName.length());
   fileType = Server1.MimeTypes.getProperty(fileType);
   header = "HTTP/1.0 200 OK\n" +
       "Allow: GET\nMIME-Version: 1.0\n"+
       "Server : HttpServer : a Java Local HTTP Server\n"+
       "Content-Type: " + fileType + "\n"+
       "Content-Length: " + fileLength +
       "\n\n";
   }
}
```

Listing 5-3

Continued

Parsing the query string seemed a natural place to use a *Hashtable*. A *Hashtable* object is—for practical purposes—an associative array. So given a *Hashtable* named *vars* and a URL of

```
http://tini/somefile.htm?Name=Al&Member=Y
```

you'd like to be able to write code like this:

```
if (vars.get("Member").equals("Y")) memberpage();
```

This is a simple matter of working through the string searching for ampersands and equal signs. The URL decoding occurs after parsing so that any equal signs or ampersands in the data remain encoded and don't confuse the parsing code.

Once the program populates the *Hashtable*, it calls the *action* method. The *action* method requires two arguments: the filename, and a *Hashtable* with the query string variables in it. The function returns the filename that the server should deliver to the client.

The *action* function in Listing 5-3 simply prints the contents of both arguments to the Java console and returns the same filename that the server passed to it. However, classes that extend the server class can provide more meaningful implementations of *action*.

The final change is to filter each character the server sends to the browser (this is in the *SendFile* class). If the server detects a backquote character, it examines the next character before proceeding. If the next character is a backquote, the server sends a single backquote to the client. Any other character is passed to the *tiniTranslate*. Along with the character, *tiniTranslate* receives a *DataOutputStream* object that it can use to send data to the browser. In this way, the *tiniTranslate* function can replace placeholders (that is, backquotes followed by another character) with any sequence you like. The example returns the time in response to 'T (using the *getTime* method which is very similar to *getContent* in the last example), but you can easily override *tiniTranslate* to do any function you like.

With these changes made, the class becomes a useful base class that you can extend to make customized servers with very little work. Just extend the server class, and provide your own *action* call.

More Open Source

Webster is attractive because it is short, sweet, and free. However, there is another free Web server out there that allows you to extend it using servlets. Servlets are to Web servers what applets are to Web browsers—small bits of Java code you write that become part of the main program (in this case, the Web server).

Servlets take advantage of Java's dynamic loading capabilities. When you use servlets, the Java VM runs and loads classes as you need them. Once you load a class, you don't need to reload it unless it changes.

So why not use applets instead of servlets? Applets will always have their place, of course. But there are many things applets can't do (or can't do easily). For example, what if you want to track the number of hits your site receives? An applet is not a good choice for that. What if you want to log those accesses on a server-side database? Again, an applet that could do that would be very complex, requiring a network connection back to the server.

Another reason you might want to avoid applets is that only certain browsers support them. If you use an applet to do something, will it run on Lynx (a text-based browser)? No. Lynx doesn't support applets. Some users that have applet-capable browsers disable applets for speed or security treasons. With servlets, nearly any browser can use your pages. The only software you'll need is on the server, where you have total control over the software you install.

Servlet Basics

To get started with servlets you should download the Java Servlet Development Kit (JSDK) from http://java.sun.com/products/servlet/index.html. Before you tackle the TINI, you might want to experiment with the JSDK on the workstation, where debugging is easier.

If you can get one of the example servlets from the JSDK running, you're ready to make your own servlets. (If you can't, then you have a configuration problem that you must fix before you can proceed.) These are just like any other Java program, but they do import some classes from the JSDK, so you'll need the servclasses.zip file mentioned in your *classpath*.

Just as an applet extends the *Applet* class, a servlet extends the *HttpServlet* class. There are three methods you usually want to override in this class: *service, init,* and *destroy.* You'll nearly always override *service,* because this is the function the Web server calls to ask the servlet to perform its designated action. The server also calls *init* when it first loads the servlet, and calls *destroy* when the servlet shuts down. You can also provide a *getServletInfo* function that returns a string with the servlet name and version number.

The *service* method receives two arguments. One (of type *ServletRequest*) lets you read the Web browser's headers and other information related to the request. The other argument (a *ServletResponse* object) is used to set the server's response to the browser. The *init* function also receives an argument (of type *ServletConfig*) that lets you read any parameters or startup configuration for the servlet.

From the user's perspective, there are several ways to see a servlet in action depending on the server in use. The TINI server—like most servers—allows you to invoke a servlet using a special pseudopath and the class name. This is similar to asking for a common gateway interface (CGI) script by entering its URL or by clicking on a link that refers to it. The *service* method (or one of the functions it calls) supplies the entire resulting Web page. It can also redirect the request to another URL or take any other action that a normal CGI script might.

Writing Your First Servlet

I'd bet just about everyone writes their first servlet by modifying one of the example servlets from the JSDK. You can practice on your usual workstation before you try to set up the TINI. There are several free servlet-capable Web servers including Tomcat (http://java.sun.com/products/jsp/tomcat/), JRun (http://www.allaire.com/), and ServletExec (http://www.servletexec.com/index.jsp). The easiest thing to do is to write out some data with the response object.

Servlet architecture is simple enough. Each servlet extends *HttpServlet*, which provides all the basic requirements. The *service* method receives a *request* parameter and a *response* parameter. The servlet can use *request* to learn about the Web browser and the document it wants. The *response* parameter allows the servlet to send data back to the browser.

For a simple servlet, you might want to simply override *service*. However, for more complex applications, you'll want to let the default *service* routine execute. It then calls different functions depending on the type of request received. For example, an HTTP *GET* request calls *doGet* and a *POST* generates a call to *doPost*. As you'd expect, each of these functions receives the same *request* and *response* parameters. If you don't override *service* and you don't override a specific request's function (like *doGet* for *GET*), then that request type will generate an HTTP Bad Request error.

You'll find a simple example of a test servlet in Listing 5-4. Notice that the *response* object argument allows you to set the type (via *setContent-*

Type), and retrieve an output stream you can use to write to the HTML output (*getOutputStream*). Once you have the output stream, you can use *println*, for example, to write output that the user will see. You can find an overview of the methods in *HttpServlet, HttpServletRequest,* and *HttpServletResponse* in Tables 5-1, 5-2, and 5-3, respectively.

```java
import java.io.IOException;
import java.io.PrintWriter;
import java.util.Date;
import javax.servlet.*;
import javax.servlet.http.*;
public class AWCTest extends HttpServlet
{
    public void doPost( HttpServletRequest req, HttpServletResponse res)
    throws ServletException, IOException
    {
        doGet(req, res);
    }
    public void doGet( HttpServletRequest req, HttpServletResponse res)
    throws ServletException, IOException
    {
        // Do HTTP header stuff
        res.setContentType( "text/html");
        // Get Writer
        PrintWriter pw = res.getWriter();
        // Start HTML
        pw.println( "<html>");
        pw.println( "<head>");
        sendHead(pw);
        pw.println("</head>");
        pw.println( "<body>");
        // Send HTML body
        sendBody( pw);
        // Finish HTML
        pw.println( "</body></html>");
```

Listing 5-4
Example of a Test Servlet

```
        }
// Send Head section
  public void sendHead(PrintWriter pw) {
    pw.println("<title>AWC Servlet</title>");
  }
// Send Body section
  public void sendBody( PrintWriter pw)
  {
    pw.println( "<H1>Test!</H1>");
        pw.println(getCurrentTime());
  }
public static String getCurrentTime()
{
        return new Date().toString();
}
}
```

Listing 5-4
Continued

	Method	Description
TABLE 5-1	**Method**	**Description**
The *HttpServlet* Class	doDelete	Handles an *HTTP DELETE* request
	doGet	Specifies code to execute when a *GET* request arrives; you'll nearly always override this method
	doHead	Handles the *HTTP HEAD* request
	doOptions	Executes in response to an *OPTIONS* request; rarely overridden since the default method does the necessary work
	doPost	Handles an *HTTP POST* request; you'll often override this if you are handling form data
	doPut	Executes when a *PUT* request arrives
	doTrace	Handles a *TRACE* request; the default properly responds, so you'll rarely override this method
	getLastModified	Returns time last modified; the default method will work, but if you can provide a more accurate time browser, caches and proxies will work more efficiently
	service	Receives raw requests and dispatches them to *doGet, doPut,* etc.; might override this if you need to handle custom commands

TABLE 5-2	**Method**	**Description**
The *HttpServletRequest* Class	getAuthType	Returns the name of authorization type, if any
	getContextPath	Returns the context path of the URL
	getCookies	Returns an array of cookies (cookies are data stored for the server by the browser)
	getDateHeader	Returns a header value as a date (long) value
	getHeader	Returns the specified header as a string
	getHeaderNames	Returns an enumeration that lets you iterate the request's header names
	getHeaders	Returns an enumeration that lets you iterate the request's header values
	getIntHeader	Returns a header value as an integer
	getMethod	Returns the name of the method (*GET* or *POST,* for example) as a string
	getPathInfo	Returns extra path information if any (this is the portion after the servlet name, but before the query string)
	getPathTranslated	Same as *getPathInfo,* but also translates the virtual path to a real path
	getQueryString	Returns query string (the portion of the URL after the ?)
	getRemoteUser	Returns the user's name, if known
	getRequestedSessionId	Servlets use a session ID to store state information; each request may provide a session ID to associate itself with a previous request, but this ID may be in error or may be expired
	getRequestURI	Returns request portion of URL
	getRequestURL	Reconstructs entire URL (except for query string)
	getServletPath	Returns name or path of servlet
	getSession	Returns (and possibly creates) an *HttpSession* object that tracks this request's session
	getUserPrincipal	Returns a security principal object associated with the user
	isRequestedSessionIdFromCookie	Determines if session ID is from cookie
	isRequestedSessionIdFromURL	Determines if session ID is from URL
	isRequestedSessionIdValid	Determines if requested session ID is valid
	isUserInRole	Determines if user is authorized for given role

TABLE 5.3

The *HttpServletResponse* Class

Method	Description
addCookie	Adds a cookie to the response
addDateHeader	Adds a header with a date value
addHeader	Adds a string header
addIntHeader	Adds an integer header
containsHeader	Determines if a header is already present
encodeRedirectURL	Encodes a URL for use with *sendRedirect*
encodeURL	Encodes a URL and includes a session if necessary
sendError	Sends an error to the browser
sendRedirect	Sends a redirect request to the browser
setDateHeader	Sets an existing header with a date value
setHeader	Sets an existing header with a string value
setIntHeader	Sets an existing header with an integer value
setStatus	Sets the status code (for example, 404 is page not found) to return

TINI Servlets

Once you have a good idea how servlets work, how do you put them on the TINI? First, you need the TINI servlet Web server from http://www.smartsc.com/tini/TiniHttpServer. This is in addition to the JSDK and, of course, the normal Java and TINI software.

The server (written by Smart Software Consulting and released under the GNU General Public License) is very easy to install. You'll want to follow the directions included with the download since you may have a newer version of the software than I used. However, the general steps are simple:

Unzip the distribution file, making sure to preserve the directory structure (in other words, allow the unzip program to create new subdirectories that the zip file contains).

Edit deploy.bat to reflect your TINI root ID and password. If you haven't changed them from the default (root and tini), you can skip this step.

Run deploy.bat passing the IP address or name of your TINI.

Log into your TINI and enter: source /bin/TiniHttpServer. This will start the server and return you to the command prompt. The server takes a bit of memory, so you may want to issue a *few gc* commands before you start. Also, make sure you don't have any other Web servers running on the TINI and clean up any unnecessary files.

Now you should find the server's home page by browsing to your TINI.

You can add more documents (or replace existing ones) in the /docs directory of the TINI.

The server's main page points to several servlets you can try that are built into the server by default. You should view these and their source code to get a feel for how the TINI server works.

Building for TINI

As usual, the TINI's build procedure is a bit different. Luckily, the *Tini-HttpServer* installation has batch files that simplify things considerably. Again, follow the directions included with the server, but here are the basic steps:

Modify tinienv.bat. Do this to reflect the directories you use for TINI, the JDK, and the JDSK.

Run the onetime.bat file. This builds several things that the server needs. As the name implies, you only need to do this step one time.

Edit compile.bat. Within this file, you'll see a line that sets the variable *SERVLET_SRC.* This line contains a list of java files that correspond to the servlets the server will use. Add your source file name here, and also remove any of the files you don't want in your finished server. If your source file is in the src directory, you can just name it; otherwise, use a path relative to src, or an absolute path. If you remove a servlet from this line, also remove its class file from the classes.tini subdirectory.

Run compile.bat. This builds the server with the selected servlets.

Run tiniconv.bat. This builds the tini file you will execute.

Run deploy.bat again. Remember to specify the host address or IP address of your TINI board as an argument to the batch file. This will copy all the files to the TINI.

Now you should be able to start (or restart) the server as before and browse to your servlet. Of course, if you encounter any compile or conversion errors, you'll need to fix them before you can proceed.

Configuration

The *TiniHttpServer* application allows you to set options in its property file. The property file is, by default, located at /etc/server.props (you specify this file name on the server's command line). You can alter many of the operating parameters using this file (see Table 5-4). In addition, the file specifies a servlet property file (by default, /etc/servlets.prop). By altering this file (see Table 5-5), you can provide aliases for servlets, set servlet arguments, and preload any servlets to avoid the overhead of loading them on the first request.

TABLE 5-4

Configuration Options for *Tini-HttpServer*

Property name	Default value	Purpose
server.bufferSize	512	Size of the input and output buffers (in bytes).
server.docRoot	/docs	Specifies the document root directory of *Tini-HttpServer.*
server.indexFile	index.html	Specifies the name of the file to show if the requested URL specifies just a directory.
server.logFile	(No default)	Specifies the name of the log file. If not specified, *TiniHttpServer* will send all output to *System.out.*
server.mail.from	(No default)	Specifies the text to use in the *From:* header when e-mailing log files.
server.mail.to	(No default)	E-mail address to which log files will be e-mailed when they reach the size specified by *server.maxLogSize.*
server.maxLogSize	(No default)	Specifies the maximum size that the server log or transfer log can attain before it is mailed to the *server.mail.to* address.
server.maxHandlers	5	Specifies the maximum number of concurrently handled requests.
server.mimeTypesFile	(No default)	Specifies the name of a properties file containing additional extension to MIME-type mappings. The lines should be formatted *.ext= mime/type.*
server.port	80	Specifies the TCP/IP port on which *Tini-HttpServer* listens for requests.
server.requestTimeout	10	Specifies the time (in seconds) the server will wait for a request. If a browser connects but does not send a request within this time, *TiniHttpServer* will respond with *408 Request-timeout.*

	Property name	Default value	Purpose
TABLE 5-4 Continued	server.stackTrace	False	Specifies whether to print a stack trace when logging exceptions. Set to *true* to enable.
	server.transferLog	(Server log)	Specifies the file which holds transfer log entries. If omitted, the entries appear in the server log. Set to "-" to force entries to *System.out*.
	server.verbose	False	Specifies whether to display properties on start-up; set to *true* to enable. This property can also be enabled by specifying *-v* on the command line.
	servlet.prefix	/servlet/	Specifies the URL prefix which signifies that this request is intended for a servlet.
	servlet.propFile	/etc/servlets.props	Specifies the name of the servlet properties file.
	session.name	sscSessionId	Specifies the name used for the automatic session tracking cookie or URL parameter (depending on which session type is configured).
	session.timeout	3600	Specifies the initial number of seconds that are allowed between requests for a particular session. Sessions that exceed this time between requests will be invalidated and discarded.
	session.type	Cookie	Specifies the technique used for automatic session tracking. Set to *Cookie* to use cookies. *URL* to use URL rewriting, or *None* to disable automatic session tracking.
TABLE 5-5 Servlet Property File Options	<alias>.code	N/A	Creates an alias for the servlet class specified by the property value. If your servlet class is *com.acme.BigMagnetServlet,* then adding the line *magnet.code= com.acme.BigMagnetServlet* to your servlet properties file will allow users to access your servlet as http://kumquat/servlet/magnet instead of http://kumquat/servlet/com.acme.BigMagnet-Servlet (assuming your TINI is named *kumquat*).
	<servlet.class.name>.initArgs	N/A	Specifies initial arguments for a servlet. The property value contains the initial arguments as name=value pairs separated by semicolons.
	<servlet.class.name>.preload	N/A	If set to *true,* the servlet will be preloaded when *TiniHttpServer* starts. If set to *false,* the servlet will not be loaded until the first request for it is received.

A Tini Servlet

Listing 5-5 Shows the AWC applet. This simple applet just retrieves the time and other information from the TINI, but it would be just as simple to use the servlet to control outputs or display the state of inputs.

Look at the listing carefully. The *doGet* function will be the same most of the time. First, it sets the content type. Next, it retrieves a *PrintWriter*

```java
import java.io.IOException;
import java.io.PrintWriter;
import java.util.Date;
import javax.servlet.*;
import javax.servlet.http.*;
//import com.dalsemi.system.Clock;
import com.dalsemi.system.TINIOS;
public class AWC extends HttpServlet
{
    public void doPost( HttpServletRequest req, HttpServletResponse res)
    throws ServletException, IOException
    {
        doGet( req, res);
    }
    public void doGet( HttpServletRequest req, HttpServletResponse res)
    throws ServletException, IOException
    {
        // Do HTTP header stuff
        res.setContentType( "text/html");
        // Get Writer
        PrintWriter pw = res.getWriter();
        // Start HTML
        pw.println( "<html>");
        pw.println( "<head>");
        sendHead(pw);
        pw.println("</head>");
        pw.println( "<body>");
        // Send HTML body
```

Listing 5-5
AWC Applet

```
            sendBody( pw);
            // Finish HTML
            pw.println( "</body></html>");
    }
// Send Head section
    public void sendHead(PrintWriter pw) {
      pw.println("<title>AWC Servlet</title>");
    }
// Send Body section
    public void sendBody( PrintWriter pw)
    {
            pw.println( "<H1>TINI Statistics</H1>");
            pw.print  ( "This page was generated by AWC at <b>");
            pw.print  ( getCurrentTime());
            pw.println( "</b>.<p>");
            pw.print  ( "TINI OS Firmware Version is <b>");
            pw.print  ( TINIOS.getTINIOSFirmwareVersion());
            pw.println( "</b>.<p>");
            long uptime = TINIOS.uptimeMillis() / 1000;
            long uptimeHours = uptime / (60 * 60);
            long uptimeMinutes = (uptime % (60 * 60)) / 60;
            long uptimeSeconds = uptime % 60;
            pw.print  ( "This TINI has been up for <b>");
            pw.print  ( uptime);
            pw.print  ( " seconds (");
            pw.print  ( uptimeHours);
            pw.print  ( " hours, ");
            pw.print  ( uptimeMinutes);
            pw.print  ( " minutes, ");
            pw.print  ( uptimeSeconds);
            pw.print  ( " seconds)</b>.<p>");
    }
    public static String getCurrentTime()
    {
            return new Date().toString();
    }
}
```

Listing 5-5

Continued

object that corresponds with the HTML output stream. Then it writes the Web document. To make things a little more general-purpose, the *doGet* method calls *sendHead* and *sendBody* to fill in the appropriate sections of the HTML document. That makes it easy to extend the class and customize the output.

The *doPost* method simply calls *doGet* so that the same logic handles both cases. If you want to know more about the URL, the query string, or a form submission, you can read the data from the *HttpServletRequest* object. If you want more control over the output (for example, a custom return code), you can use the *HttpServletResponse* object.

As with most TINI code, once you know how it works in Java, you know how it works with the TINI. That also means you can debug your code—at least most of it—on nearly any Java platform that supports servlets.

What about Applets?

Another Java-based technology you often hear about in conjunction with the Web is applets. However, for the TINI, applets aren't terribly useful. Still, there may be times you'll want to build an applet that communicates with the TINI and that can present a few problems.

Applets are Java classes that extend a special Java class known as *Applet*. Other programs (especially Web browsers) can load these classes and display them as though they were built into the program. This allows a Web page to show—for example—an analog clock that shows the current time. To do this, the applet must draw on a *Graphics* object—something that most Java books talk about in some detail but this one does not. The Web browser or other host program incorporates the applet's drawing into its display. In the case of the Web browser, it appears that the Web page includes interactive, animated objects.

There are many reasons applets can be troublesome. First, not all browsers support applets (Internet Explorer and Netscape Navigator do, but many other browsers do not). Second, even if the browser supports applets, the user may disable them.

The third problem that may arise is that a malicious applet could damage the user's system. To prevent this, most browsers hobble applets by default. For example, an applet can't read or write to the user's hard drive. The applet can't connect to any network sockets other than back to the host that sent the applet in the first place. It is possible for the user to override these security lockouts, but it is unusual to find anyone who does. You can also sign an applet with a digital signature (which you must purchase from one of a handful of companies that issue signature certificates). If your applet is signed, and if the user trusts your signature, your applet may be able to do some things it could not otherwise. However, there is no assurance you'll be able to do anything at all. To make matters worse, there is no single way to sign an applet for both commonly used browsers.

So what does this mean for the TINI? Practically nothing. No one is going to browse the Web with a TINI, so you'll never need to write an applet for the TINI. However, you might want to create an applet that talks back to the TINI. However, the applet can only open a network connection to its original server, which implies that the TINI must serve the applet (which is just an ordinary class file or a jar file). You'd build this applet on the PC and simply store it on the TINI. It can't run on the TINI, and if you build it for the TINI, it won't run on the workstation.

An Applet Example

Applets can be as simple as you want them to be. Obviously, if you want to write a large user interface, this is going to be difficult. However, a large user interface will be difficult in any language—it isn't the applet's fault. To create an applet, you extend the *java.applet.Applet* object. There are seven methods you might want to override to provide your applet's behavior:

init. This provides code the applet executes when the browser first loads your code.

destroy. The final call the browser makes to the applet.

start. The browser calls *start* to inform your applet it is visible.

stop. The browser calls *stop* when the applet is invisible. For example, the page the applet appears on scrolled away from the applet.

paint. Use this to draw the applet's contents.

getAppletInfo. Returns optional information about the applet (used by some tools to show an about box for the applet).

getParameterInfo. Returns optional information about the applet's parameters.

At first glance, it seems that *init* and *start* (along with *destroy* and *stop*) are redundant. However, the difference is that *init* (along with *destroy*) occurs exactly one time. The *start* and *stop* calls may occur frequently to notify you that your applet is visible or not. This is useful for applets that perform a lot of processing. If they aren't visible, they can suspend that processing and essentially wait in a dormant state until a call to *start*.

Drawing

The *paint* function is the one that will be least familiar to the embedded programmer. This function sends a *Graphics* object that you can use to draw on the face of the applet. Even writing text is really drawing.

Listing 5-6 shows a very simple applet that just draws some text. Notice the only method override is *paint*. The *drawString* method prints the text, and all the other methods rely on the default implementations in the *Applet* class.

The Web page to display the applet appears in Listing 5-7. You can optionally supply parameters to the applet that it can read during its processing. In this case, you don't need any parameters, but you could read them using the *getParameter* method.

You can serve this applet off a TINI (or any Web host, for that matter), but that's not really making use of the TINI. However, if you want to

Listing 5-6
Simple Applet That
Draws Some Text

```
import java.applet.*;
import java.awt.*;
public class TestApplet extends Applet {
 String value="Test Message";
 public void paint(Graphics g) {
    g.drawString(value,10,10);
 }
}
```

Listing 5-7
Web Page to Display
the Applet

```
<APPLET CODE=TestApplet.class HEIGHT=60 WIDTH=150>
</APPLET>
```

make a network connection to the TINI, you'll have to serve the applet from the TINI. That's because most browsers will not allow applets to connect to any server except the one that sent it. This prevents a malicious applet from, for example, sending e-mail using your computer with no way to trace who was responsible.

Networking

Luckily, it is a simple matter to build an applet with a socket connection back to the TINI. Of course, then you need a corresponding socket running on the TINI. You could make the applet socket request something from the Web server or FTP server that is already running. However, it is easy enough to write a customized server, as you've seen in previous examples.

Although the graphics part of an applet may be new, the networking is the same as you've seen before. Look at the applet in Listing 5-8. It

```
// Switch applet
import java.applet.*;
import java.awt.*;
import java.net.*;
import java.io.*;
public class SwApplet extends Applet implements Runnable {
 String value="???";
 public void start() {
     new Thread(this).start();
 }
 public void run() {
     Socket sock;
     try {
         sock=new Socket("TINI",111);
         PrintWriter pw=new PrintWriter(new OutputStreamWriter(sock.getOutputStream()),true);
         BufferedReader in=new BufferedReader(new InputStreamReader(sock.getInputStream()));
```

Listing 5-8
Applet Listing

```
        while (true) {
            pw.println("?");
            String rvalue=in.readLine();
            if (rvalue.charAt(0)=='0') value="Off"; else value="On";
            repaint();
            try {
                Thread.sleep(1000);
            } catch (Exception e) {}
        }
    } catch (Exception e) {value="Err"; repaint(); }
}
public void paint(Graphics g) {
    g.drawString(value,10,10);
}
}
```

Listing 5-8
Continued

communicates with the example server from Chap. 4. The networking portion is the same as the corresponding client in that chapter. In this case, the applet (which runs a separate thread) queries the server repeatedly, pausing for 1 s between requests.

All the details required to write applets are beyond the scope of this book. However, you can easily add controls like buttons and text fields to applets without drawing. You can find plenty of information about applets in nearly any Java book, or—of course—on the Web.

SUMMARY

The TINI really shines as a Web server. Sure, you can spend your time writing custom client interfaces, but why not leverage all the work Microsoft, Netscape, and others are spending on Web browsers? Use their interfaces and you can quickly and inexpensively have a cross-platform solution.

The only downside to using the TINI as a Web server is that it is easiest if you have an Ethernet connection available. While it is possible to interface the TINI to a modem and dial an Internet service provider (ISP), it isn't nearly as easy as using the built-in Ethernet.

Of course, many buildings now have high-speed Ethernet already available. Besides that, many high-speed modems (DSL and cable modems, for example) appear as Ethernet ports. As a last resort, you can always tie the TINI (or a lot of TINIs) to a PC gateway that connects to the Internet.

Java Jr.

You may have noticed that the TINI is like a miniature PC. The Java is practically the same as workstation Java, and you have similar resources (although there is, of course, less memory and processing speed, as you'd expect). However, the TINI is not—by any stretch of the imagination—a classic embedded computer. Typical embedded computers have many I/O commands and are usually geared toward performing tasks in a predictable amount of time.

One simple popular embedded computer is the Parallax Basic Stamp. It has powerful commands that can create and measure pulses, generate pulse width modulation, and perform other sophisticated I/O tasks. To address the Internet market, Parallax introduced a new chip—the Javelin Stamp—that is similar to the Stamp, but uses a Java-like language and has special features to facilitate networking as well as other modern programming techniques that were not possible with the Basic Stamp.

Like the TINI, you should understand how the Javelin Stamp works before you tackle networking, so this chapter will show you how to use the Javelin Stamp in a simple application. You might wonder which is better, TINI or Javelin Stamp? That's like asking what's better, a hammer or a screwdriver? Both have strengths and weaknesses. The Javelin Stamp has many benefits:

- Easy connection to the outside world with 16 I/O pins and flexible I/O commands
- Deterministic timing—no multithreading or garbage collection issues
- Easy-to-use 24-pin DIP form factor

On the other hand, compared to the TINI, there are some limitations:

- There is no built-in Ethernet port.
- There is no multitasking or garbage collection (this is a mixed blessing; see above).
- Javelin Stamp's Java is different in many ways from normal Java.

Hardware

The Javelin Stamp is a 24-pin device that resembles an IC (see Fig. 6-1). Sixteen of the pins (pins 5 to 20) on this chip are for input and output devices (you can configure each pin in your program). Of the eight remaining pins, one (RES on pin 22) is normally not connected to any-

Figure 6-1
The Javelin Stamp

thing (unless you want an external reset switch, which is often not necessary). Pins 1 to 4 form an RS232 port for connection to a PC. You'll use this connection to program the Javelin Stamp.

The three pins remaining are for power connections. The Javelin Stamp has its own voltage regulator, so you can connect an unregulated power supply directly to pin 24. The onboard voltage regulator will convert the input voltage to 5 V and it provides 5 V on pin 21 (of course, there is a limit to how much current you can draw from this pin; consult the manual). Alternatively, if you have a 5-V regulated supply, you may connect it directly to pin 21 and leave pin 24 not connected. Whether you connect pin 21 or pin 24, always ground pin 22.

Parallax offers a special board that allows you to easily connect the chip to power and a PC. This board is convenient, but not strictly necessary— you can make the appropriate connections yourself using a solderless breadboard or your own custom circuit.

Programming

The Javelin Stamp uses a special Integrated Development Environment (IDE) that runs on your PC. The IDE allows you to write programs, download programs to the Javelin Stamp, and also debug running programs.

Run the IDE program from the Start menu. You'll see an initial window like the one in Fig. 6-2. The first step is to make sure everything is working properly. Enter a simple Java program like the one in Listing 6-1.

Use the Save option on the File menu to save the file to a directory of your choice. Of course, the file name must be JStampTest.java.

Running the Test Program

If you haven't run the IDE program before, you'll have to tell it which serial port you have connected to the Javelin Stamp. Select the Options item from the Javelin Stamp menu. A form will appear with several tabs—select the Debugger tab. You'll see a screen like the one that appears in Fig. 6-3. If you know the serial port you are using, you can select it from the drop-down list. You can also click the Auto button.

Figure 6-2

The IDE Allows You to Write, Build, and Debug Programs

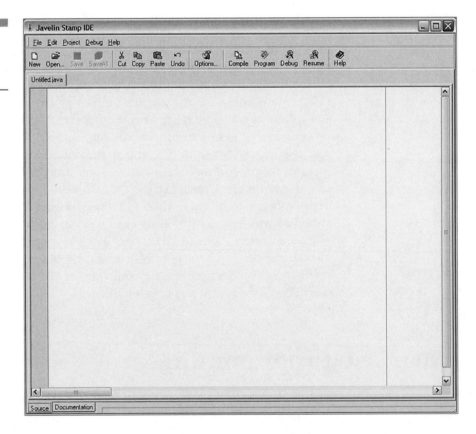

Listing 6-1
Simple Java Program
to Test Javelin Stamp

```
public class JStampTest
  {
  public static void main()
    {
      System.out.println("Javelin Stamp Test");
    }
  }
```

Figure 6-3
The Javelin Debug
Window

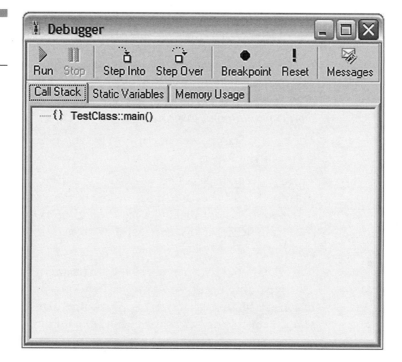

This will cause Javelin Integrated Development Environment (JIDE) to search for the Javelin Stamp automatically.

When you have the correct serial port selected, click OK to close the option form. Next, select the Debug item from the Javelin Stamp menu (or press <*Control+D*>). You should see a progress indicator as JIDE compiles your code and sends it to the Javelin Stamp. When the download is complete, you'll see a green arrow next to the first line of your program (which will appear with a green background). In addition, a Debug window will appear (see Fig. 6-3 above). In the Debug window you'll see a green Play button. Click on this button (or press <*F9*>) and your program will execute. To see the message, click on Debug | Show Message Window in the JIDE main menu. You'll see the text you entered into your program in the message window (Fig. 6-3). The Javelin Stamp sent this string to your PC via the serial port.

Javelin Stamp Architecture

The Javelin Stamp's language is practically Java (in fact, like the TINI, it uses an ordinary Java compiler). There are some very distinct differences between the two in the runtime interpretation. Here's the short list:

- The *int* type is 16 bits wide. Therefore, the largest value you can place in an *int* is 32,767 (0x7FFF). The *byte* type is 8 bits, and the *short* type is 16 bits, just as in regular Java.

- The Javelin Stamp does not allow floating point types (*float* and *double*) or the *long* type.

- There is no garbage collection. Once you allocate memory, it is not reclaimed until the Javelin Stamp resets. For embedded systems, however, this is not usually a big limitation, as opposed to garbage collection, which can wreak havoc with real-time system performance.

- Not all Java libraries are available. Others are different (because of data type differences, for example). In addition, the Javelin Stamp has many libraries that don't appear in regular Java. These allow you to control the hardware and peripheral devices.

- The Javelin Stamp only supports one thread.

- Javelin Stamp strings and characters are composed of ASCII characters, not Unicode characters.

- Interfaces are not available.

- The Javelin Stamp only supports single-dimensioned arrays.

At first, these differences seem a bit draconian, but in fact, for the jobs you want the Javelin Stamp to do, they aren't very restrictive at all. Even better, the lack of garbage collection and threads prevents nasty timing issues that invariably arise when you are trying to perform low-level programming.

Of course, threads are very useful in some cases. Luckily, the Javelin Stamp has a workaround that allows you to perform several important functions in the background. These are what the Javelin Stamp calls *virtual peripherals*.

The Javelin Stamp has two broad categories of programming commands. First, there are ordinary objects that allow you to perform common operations (for example, you can set the state of an output pin or read an input pin). However, some objects are special—they are the virtual peripherals, or VPs for short.

The Javelin Stamp reserves several slots for VPs. Periodically, the Javelin Stamp stops what it is doing and executes each active VP. VPs are associated with objects like *Uart*, which performs serial communications. Since they run in the background, they can do things like receive serial data without impeding your main program.

Of course, VPs must be small and fast, and they have to be built into the Javelin Stamp's firmware, so you can't write your own. Also, you can only run a certain number of VPs at once, so they are not as flexible as threads.

Javelin Stamp I/O

The key to understanding the Javelin Stamp's I/O system is the *CPU* class. This static class allows you to manipulate each pin and perform other useful I/O tasks. For example, you can generate (or measure) pulses. Other objects include *EEPROM* for storing data in nonvolatile storage and the VPs *Uart, Pwm,* and *Timer.*

Digital I/O uses the *readPin* and *writePin* methods of the *CPU* class. If you'd like to read pin 0, for example, and copy its output to pin 1, you'd write:

```
CPU.writePin(CPU.pin1,CPU.readPin(CPU.pin0));
```

The second argument to *writePin* and the output of *readPin* are both *boolean* data. Don't try to use 0 and 1 instead of *pin0* and *pin1*—it won't work (*pin0* is actually 1, *pin1* is really 2, and *pin15* is equal to 0x8000).

In addition to normal digital I/O, you can take advantage of the *CPU* object's special methods to perform advanced tasks you'd normally have to write software to handle. For example, the *count* method counts the number of rising or falling edges on a pin over a specified interval. Table 6-1 shows the most important methods you can use. You'll find more details about them and other methods in the Javelin Stamp documentation.

In addition to *CPU,* there are other special Javelin Stamp classes that handle other embedded system tasks, for example, the *Button* object

Other than the limitations mentioned earlier in this chapter, and the special classes like *CPU,* using the Javelin Stamp is just like using any other Java programming environment. For example, Listing 6-2 shows the classic blinking LED program adapted for the Javelin Stamp.

The class name, *LEDDemo,* can be anything you like. However, the Javelin Stamp will look for a *main* subroutine (which must be static). Unlike conventional Java, *main* takes no arguments since there isn't really

TABLE 6-1

CPU I/O Methods

Method	Description
count	Counts edges on a pin over a specified time.
pulseIn	Measures a pulse width.
pulseOut	Generates a pulse of a specified width.
rcTime	Measures the charge or discharge of a resistor/capacitor (RC) network. This allows you to measure a resistance or capacitance (for example, a thermistor or an analog joystick).
readPin	Reads a digital input.
shiftIn	Reads data from synchronous devices.
shiftOut	Writes data to synchronous devices.
writePin	Sets the state of a digital output.

Listing 6-2
A Blinking LED

```
import jstamp.core.*;
public class LEDDemo {
 public static void main() {
// Set LED initial state
  boolean ledstate=false;
  while (true) {
   CPU.writePin(CPU.pin1,ledstate);
   ledstate=!ledstate;
   CPU.delay(10000);  // 1 second
   }
  }
}
```

a command line for the Javelin Stamp. If you do include arguments to *main,* that makes it a different function and the chip will not find it.

The only other part of Listing 6-2 that bears explanation is the *delay* subroutine. You'll notice that the value to delay for 1 s is 10000. You can intuit from this that the delay period is 100 μs.

Running a Program

The Javelin Stamp includes a development environment that incorporates a text editor, a help system, the program downloader, and a debugger. To run the program in Listing 6-2, you'll simply enter the text into the text editor, save it, and then select the Run button on the toolbar. This will compile the program, link it, and download the program to the Javelin Stamp. The Javelin Stamp will execute the program. Also, when the Javelin Stamp restarts—even if the PC is not connected—it will execute this same program (until you download another one, of course).

You can also use the debugging portion of the environment to set breakpoints, view messages, and step through your code. The Javelin Stamp only holds one program at a time, so if you load a different program to debug, you lose the original program (until, of course, you reload it).

Virtual Peripherals

One thing the Javelin Stamp lacks is threads. At first, this would seem to be a serious limitation since many embedded systems have to perform tasks in the background, or won't want to block to perform certain operations. However, threads also make it difficult to predict the absolute response of a Java program, which is why the Javelin Stamp does not support them.

What the Javelin Stamp does provide are virtual peripherals. These are modules provided by Parallax (you can't create your own) that install interrupt handlers to perform a variety of tasks. The Javelin Stamp has a periodic interrupt that handles its various timing functions. There are several "slots" reserved for VPs that the user can install. You can install up to six VPs.

For example, the *Uart* object is an example of a VP. Once installed, it handles serial input and output. Since it is interrupt-driven, the *Uart* provides a buffer for your program. As far as you can tell, the *Uart* class is just an ordinary Javelin Stamp object. Behind the scenes, however, it is installing an interrupt handler to do its job, and that makes it a VP.

In addition to the *Uart* object, the Javelin Stamp also provides a *Pwm* VP that generates pulse width modulation output. A PWM output can control a motor speed or a lamp brightness, or—with an additional RC network—develop an analog voltage. Another common VP is the *Timer* object. This object allows you to precisely code delays and other time-critical code. Although the *Timer* object installs a VP, that one VP can service any number of *Timer* objects. No matter how many *Timer* objects you create, you only consume one VP slot. This is in stark contrast to, say, the *Uart*. If you want to transmit and receive serial data, you'll need two *Uart* objects, and that will consume two of the Javelin Stamp's six VP slots.

In practice, using a VP isn't any different from using an ordinary object. For example, Listing 6-3 shows a simple example of using the *Uart* object.

Reclaiming Memory

The Javelin doesn't perform garbage collection like an ordinary JVM. This makes it easy to plan a program's timing, but it does make you more careful with object creation. To work around this, the Javelin's library contains a *Pool* class that you can extend to create pools of reusable objects. The idea is that you can make a fixed number of objects when your program starts and parcel them out to your program as needed. When you are done with an object, you return it to the pool for later reuse.

Suppose you have a class named *NetTransfer* that you will need from time to time. You decide you'll never need more than five instances of this class at one time. You might write the following:

```
public class TransferPool extends stamp.util.Pool {
 // constructor
TransferPool(int maxsize) {
 super(maxsize);
 for (int i=0;i<maxsize;i++)
  add(new NetTransfer());
 }
 // Type safe definition for checkOut
 // could reinit object here if necessary
 NetTransfer checkOutObject() {
  NetTransfer xf=(NetTransfer)super.checkOut();
  if (xf!=null) xf.clear();  // presumably .clear reinitializes the
object
  return xf;
 }
}
```

```
import jstamp.core.*;
public class ModemTest {
  final static int SERIAL_RX_PIN = CPU.pin0;
  final static int SERIAL_TX_PIN = CPU.pin1;
  final static int SERIAL_CTS_PIN = CPU.pin2;
  final static int SERIAL_RTS_PIN = CPU.pin3;
  Uart rxUart = new Uart( Uart.dirReceive, SERIAL_RX_PIN, Uart.dontInvert,
            SERIAL_CTS_PIN, Uart.invert, Uart.speed14400,
            Uart.stop1 );
  Uart txUart = new Uart( Uart.dirTransmit, SERIAL_TX_PIN, Uart.dontInvert,
            SERIAL_RTS_PIN, Uart.invert, Uart.speed14400,
            Uart.stop1 );
public void run() {
txUart.sendString("Test...\r\n");
while (true) {
  while (rxUart.byteAvailable()) {
    process((char)rxUart.receiveByte());
    }
  }
}
public static void main() {
modemTest = new ModemTest();
modemTest.run();
}
public void process(char c) {
// do something with characters here
  }
}
```

Listing 6-3
Using Uart

The *clear* method in this case reinitializes the object (otherwise, the object would have instance data set by the last part of the program that used it, if any). When the program starts, it creates the following pool:

```
TransferPool pool = new TransferPool(5); // 5 items in the pool
```

When the program needs a *TransferPool* object, you'll write code to retrieve one from the pool:

```
NetTransfer xfer=pool.checkOutObject();
If (xfer==null) ; // uh oh... nothing available
```

Finally, when you are done with the object, you'll call *pool.checkIn* passing the *xfer* instance as an argument. This does shift the onus of handling memory onto your own shoulders. However, it allows the Javelin to run timing-critical commands without worrying about garbage collection occurring at an inopportune time.

Summary

If you learned Java on a workstation, you probably didn't see much difference between the Java you know and the TINI. The Javelin Stamp, on the other hand, is quite a bit different. A casual glance might make you think that the Javelin Stamp is underpowered. This isn't the case, however.

Sure, the Javelin Stamp isn't a PC replacement—it isn't supposed to be. The Javelin Stamp's benefit is that it is simple to apply and works in a very easy-to-predict way. This is important when you are building embedded systems. For example, consider the *CPU.delay* method. With a conventional JVM running on Windows or Linux, it would be difficult to predict exactly how this method would behave. It would be relatively easy to make sure the *delay* method stopped execution for at least a certain amount of time. However, it would not be simple to set a boundary on the maximum amount of time the function would delay. After all, there might be multiple threads, garbage collection, and other considerations that would extend the time. The Javelin Stamp eliminates these uncertainties so you can easily control the processor's execution.

7

Javelin on the Internet

I like flashy sports cars. I was sorry to see DeLoreans (the famous, futuristic-looking stainless steel cars) go out of production because I always thought I might buy one. A few years after they quit making them, I saw a used one for sale and decided to check it out. What a surprise. Inside its sleek metallic interior were the guts of a Chevy Nova. Even my old Isuzu Impulse had more controls and gadgets inside. It just goes to prove the old adage that you can't judge a book by its cover.

At first glance, you might think the Javelin is the opposite of the DeLorean: all control, and no flash. While its subset of Java is good for performing control tasks, surely it can't connect to the Internet, right? Actually, the Javelin can connect to the Internet [using User Datagram Protocol (UDP) sockets] with classes provided by Parallax and a dialup Internet provider that supports Point-to-Point Protocol (PPP).

To use these classes, you don't have to know much about networking internals. However, if you need to modify them, you might need to know a little about the internal workings of PPP.

Inside Networking

Traditionally, you think of networks as having layers. Each layer on the server computer logically communicates with the same layer on the client computer. In reality, each layer communicates with the layer above and below in the networking "stack"—the layers of network software on each computer.

In the simplest interpretation of this layered approach, there are four layers. At the top is your program, of course. Under that is the layer that handles network sockets (usually either TCP or UDP). This layer talks to the next layer, which handles IP (collectively these two layers make up TCP/IP). The final layer handles the physical communications with the actual network hardware. For a PC, this might be the network card driver, for example.

The stack approach to networking has several advantages. Since each layer of the stack only talks to the layer immediately above and immediately below, your program doesn't have to worry about details like network card interfaces. The stack also helps programmers manage their programs since each layer logically talks to the same layer on other computers. Your program doesn't really know or care what magic things the lower layers add to your data. Depending on the protocol in use, your data might be split into pieces, encoded, or encrypted, and will certainly have addressing and routing information added to it. All this is transparent to you since it appears your layer in the stack (the top layer) communicates with the top layer of the remote stack. This same principle holds true of the other layers. The TCP layer doesn't have to worry about the details of the IP layer, other than that layer's external interface.

On the TINI or a PC you often want to use TCP networking since it is the most transparent protocol of all. TCP sockets create a virtual pipe that you use to send and receive data. Data you put in the pipe arrives at the remote stack, and the remote data arrives in your program.

Although this picture is idyllic, the real story is much uglier. The network may split your data into packets to better utilize network resources. It then sends each packet to the remote computer and waits for an acknowledgment. If the remote computer does not acknowledge, the stack has to resend the data, making sure that the remote won't process the data twice by accident. To improve performance, most computers will send several packets without waiting for acknowledgments for each one. That means the sending computer has to track multiple packets until they are acknowledged.

The receiving computer has a similar problem. Some packets from the remote computer might not arrive. That means packets might not arrive in order. You might get packet numbers 4, 5, and 7. Later, the remote computer might resend packet 4 (it missed your acknowledgment) and packet 6. The TCP stack is obligated to correctly reconstitute the data so that it arrives in order and with no duplicates to the top of the stack.

All of this is memory-intensive, and the Javelin doesn't have much memory. While it would be possible to set up a TCP stack that only allows one outstanding packet, there is still a lot of processing overhead involved with TCP and, for a small processor like the Javelin, UDP is often a better choice.

With UDP sockets, the sender forms a packet of data and sends it directly to another computer (or computers; you can broadcast UDP packets). The other computer doesn't acknowledge (unless you write the code to do that), and there is no assurance that packets will arrive in the order sent (or even arrive at all). Although this means you have to do a little more work, you can optimize the extra work for the limited resources of an embedded processor.

At the bottom of the stack, you normally think of an interface to a network card. However, the beauty of a stack is that the bottom layer can do anything you want. Since many Internet connections require a telephone modem, there is a standard protocol, PPP, that encapsulates TCP/IP data over standard phone modems.

PPP is the *de facto* standard for dialup Internet access and, with a little programming, it is possible to have a Javelin dial up an Internet provider, log in, and send (or receive) UDP packets to any host on the Internet. Luckily, Parallax provides classes aimed at sending UDP data over a PPP connection. In this chapter, I'll show you how to use these classes to create Internet-aware Javelin programs.

Javelin UDP Overview

There are three distinct phases involved in putting the Javelin on the Internet. First, you have to dial the modem and connect to an Internet service provider. Second, you have to negotiate a PPP connection—this is the physical layer that you use to gain connectivity for the stack. Finally, you have to set up and manage the individual protocols (IP and UDP in this case).

While this may sound daunting, the Javelin includes quite a few tools to make it easier. First, consider dialing the modem. The Javelin provides the *Uart* class that makes it simple to talk to the modem. In addition, the *stamp.util.dialer.DialerControl* object allows you to easily define the steps your program has to take to log into an ISP.

Your program simply creates a *DialerControl* object and passes it an array of objects. These objects are all part of the *stamp.util.dialer* package. Here are the objects that are available:

- *ActionSend.* Sends a string to the modem.
- *ActionWait.* Waits for a response from the modem.
- *ActionPause.* Pauses for a certain amount of time.
- *ActionResult.* Ends the script with a certain result code.

The *ActionWait* object is the most complex. It waits for a string from the modem. If the string doesn't occur within a specified period, the object provides an offset. The script will execute the step specified by the offset. If the string is found, the script continues with the next item.

This is easiest to understand by looking at a simple example. Suppose you want to send a single ATZ command to the modem and you expect the modem to respond with an OK. Here's an example script:

```
DialerAction aScript[] = {
  new ActionSend(txUart, "ATZ\r"),
  new ActionWait(rxUart, "OK", 2, -1),
  new ActionResult(ActionResult.SUCCESS)
};
```

Notice that the *ActionSend* constructor includes a carriage return in the string to send. The *ActionWait* constructor looks for an OK for 2 s. If it finds the string, the script will proceed to the next line (an *ActionResult* object that causes the script to end). If the string does not appear, the script will go back one line (−1). That will cause it to send another ATZ command.

To execute the script, you have to create a *DialerControl* object and pass the script array to its constructor. Then a call to *runScript* will execute the script commands. Here's an example:

```
DialerControl dc = new DialerControl(aScript);
dc.runScript();
```

The return value of *runScript* is the value passed to the *ActionResult* constructor for the object that ended the script's execution. Keep in mind that a script might have more than one *ActionResult* object.

Of course, a practical script will be much more complicated. For a pure PPP server, you should be able to use this script:

```
DialerAction pppScript[] = {
  new ActionPause(1000),
  new ActionSend(txUart,"+++"),        // maybe take modem offline
  new ActionWait(rxUart,"OK",3,1),
  new ActionSend(txUart, "ATZH0\r"),  // hang up and reset
  new ActionWait(rxUart, "OK", 2, 1),
  new ActionPause(3000),
  new ActionSend(txUart, "ATE0V1Q0\r"),
  new ActionWait(rxUart, "OK", 2, -1),
  new ActionSend(txUart, "ATDT8888888\r"),   // ISP Phone #
  new ActionWait(rxUart, "CONNECT", 60, -6),
  new ActionResult(ActionResult.SUCCESS)  };
```

The +++ sequence is the typical way to take a modem into command mode if it is online. Of course, you may have to adjust the script to match your modem's peculiarities, but this script will serve as a starting point. Many ISPs use an extra authentication procedure that requires you to enter a user ID and password. That's easy to handle also:

```
DialerAction loginScript[] = {
  new ActionPause(1000),
  new ActionSend(txUart, "+++"),
  new ActionWait(rxUart,"OK",3,1),
  new ActionSend(txUart, "ATZH0\r"), // hang up and reset
  new ActionWait(rxUart, "OK", 2, 1),
  new ActionPause(3000),
  new ActionSend(txUart, "ATE0V1Q0\r"),
  new ActionWait(rxUart, "OK", 2, -1),
  new ActionSend(txUart, "ATDT8888888\r"),   // ISP Phone #
  new ActionWait(rxUart, "CONNECT", 60, -6),
```

```
    new ActionSend(txUart, "\r"),
    new ActionWait(rxUart, "ogin:", 5, -1),   // wait for login:
    new ActionSend(txUart, USERNAME),
    new ActionSend(txUart, "\r"),
    new ActionWait(rxUart, "word:", 5, -3),   // wait for password
    new ActionSend(txUart, PASSWORD),
    new ActionSend(txUart, "\r"),
    new ActionResult(ActionResult.SUCCESS)
};
```

The script waits for *"ogin:"* so it doesn't matter if your ISP prompts with a *"Login:"* or a *"login:"*. The same logic applies to the password prompt. Of course, if your ISP is different, you may have to change things around a bit.

If the *"CONNECT"* string is not found, the script backs up six steps to the start. However, if the log-in or password prompts don't appear, the script only backs up to look for the log-in prompt.

Notice that the user ID and password in the above example are in variables. Some ISPs require a bit of detective work on this end as well. Most ISPs that require any unusual log-in procedure will supply a .scp file for the Windows dial up software to use when it connects. You can often figure out what is necessary by examining this file. For example, here is the .scp file for a popular ISP:

```
proc main

    string prefix = "aolnet/ent."

    waitfor
      "ANSNet"    then DoANSNetLogin,
      "Sprint-IP"     then DoSprintLogin,
      "UU.Net"    then DoUUNetLogin,
      "UQKT1"         then DoANSNetLogin,
      "UQKT2"         then DoSprintLogin,
      "Saturn.BBN"    then DoUUNetLogin
    until 20

      DoANSNetLogin:
       waitfor "login"  until 20
       if $SUCCESS then
           transmit prefix + $USERID, raw
           transmit "^M"
           goto DONE
       endif

      DoUUNetLogin:
       waitfor "login" until 20
       if $SUCCESS then
           transmit prefix + $USERID, raw
           transmit "^M"

           waitfor "Password"
           transmit $PASSWORD, raw
```

```
            transmit "^M"
            goto DONE
        endif

    DoSprintLogin:
    waitfor
        "Login"    then DoSprintConnect,
        "Username" then DoSprintConnect
    until 20

    DoSprintConnect:
     transmit prefix + $USERID, raw
     transmit "^M"

     waitfor "Password"
     transmit $PASSWORD, raw
     transmit "^M"

     waitfor "Network User"  until 5
     if $SUCCESS then
         transmit "PPP"
         transmit "^M"
     endif
     goto DONE

    DONE:
     transmit "^M"

 endproc
```

Although the programming language isn't any standard one, it is easy to work out the details by inspection. The script waits for one of several identifiers: ANSNet, Sprint-IP, UU.Net, UQKT1, UQKT2, or Saturn.BBN. Then it vectors to a different part of the script depending on what it finds.

Although you could duplicate this script using the Javelin Stamp's scripting language, in practice you probably don't need to do that much work. Instead, use an ordinary terminal program to dial up the ISP and see which identifier it sends. For the same access number, this identifier should always be the same. In my case, the dialup number returns UQKT2, a Sprint network. You can also notice if the ISP transmits *"Login:"* or *"Username"* or something else as a prompt. Sometimes script elements are optional. For example, the script shows that it waits briefly for Network User. However, in practice, the ISP never actually sends this to me, so I left it out of the corresponding Javelin script.

However, one thing you should notice about the script is that it adds a prefix to your user ID. Near the top of the script, you can see the prefix is set to *"aol/ent".* That means you'll have to use this modified user ID in the script. However, the PPP user ID doesn't use the prefix, so you'll have to specify it differently (you'll see where you specify that shortly).

So connecting to ISP via modem is made relatively simple by using the scripting objects available. You may have to do a little sleuthing to find the exact log-in steps required by your ISP, but once you do, this part of the connection is relatively simple.

One other thing to watch for is that your ISP allows PAP log-ins (or doesn't require PPP authentication at all, which could be the case if you are dialing into a dedicated server). If your ISP requires CHAP or another encoded authentication, the code presented here won't work. It would be possible, but not trivial, to add the CHAP protocol to the Javelin's log-in code.

Opening PPP

To create a network stack, you must build each layer and connect them together. For a simple Javelin program you'll use four layers: PPP, Internet Control Message Protocol (ICMP), IP, and UDP. The ICMP layer allows you to ping the Javelin. Here's a simple routine that sets up the stack:

```
// Create the IP protocol.
ip = new IP(new IPPacketPool(4));
ppp = new PPP(ip, rxUart, txUart);
// Set the physical interface.
ip.iface = ppp;

// Register the ICMP protocol.
ip.registerProcessor(new ICMP(ip), ICMP.PROTO_ICMP);

// Register the UDP protocol.
udp = new UDP(ip);
ip.registerProcessor(udp, UDP.PROTO_UDP);
```

The first line of code creates the IP part of the stack, using a pool of four packets for communications. The next line sets up the PPP layer, attaching it to the IP layer and two *Uart* objects (one for receiving and another for sending). The IP layer has a field named *iface* that contains the physical link layer and the program sets this as well. The two *registerProcessor* calls add the ICMP and the UDP protocol layers.

Once the dialer connects to the ISP, you must call *ppp.open*. If your ISP requires PPP authentication, you'll provide your PPP user name and password to this call. Otherwise, you don't need to provide anything to the call. This call will attempt to negotiate a network connection with the ISP. If it returns *false*, some error occurred. If it returns *true*, however, you may or may not have a network connection.

To allow PPP to process incoming data, you must periodically call *ppp.processInterface*. This method handles incoming packets and sends packets to the upper layers. When the link is complete, the *ppp.linkUp* method will return *true*. Until that time, the Internet connection is still not ready.

Often you'll do your processing in a loop:

```
while ( true ) {
    ppp.processInterface();
    if (  !ppp.linkUp() )
      continue;
      .
      .
      .
```

Sending and Receiving

Sending data involves creating a packet and transmitting it via UDP. Here's a snippet of code:

```
pkt=udp.startPacket(addr,64,100);
pkt.buffer[pkt.length++]=(byte)'A';
.
.
.
udp.sendPacket(pkt, true);
```

In this particular case, the program is creating a packet for a certain destination IP address (*addr*). The source port number is 64 and the destination port on the remote host will be 100. The IP address is simply a byte array containing the four elements of the address. The packet will have additional information in it, so you should never write to, for example, *pkt.buffer[0]*. Instead, use the *pkt.length* field as the start of your area in the packet and update it to indicate the new length of the packet.

The *sendPacket* call actually transmits the packet. The second argument tells the library if it should reclaim the packet when it finishes. If you want to reuse the packet, you can set this parameter to *false*. Otherwise, set it to *true* and the library will add the packet back to the pool of available packets for possible reuse later.

Often you will want to reply to an incoming packet. Obviously, that packet already has the source and destination ports, but they should be reversed. This is so common that the library provides a *startReply* method. This works just like *startPacket*, but it only requires an incoming packet. The call extracts the destination address and port, along with the source port, from the incoming packet and creates an appropriate response packet.

Receiving a packet is a bit more complex. Typically, you'll create a new class that extends *PacketProcessor*. The most important method in this class is *processPacket*. This call will receive a *Packet* object, a starting offset, and a data length. This method is your opportunity to process the data in incoming packets. Keep in mind that there may be other things before and after your data, so be sure to obey the starting offset and length arguments.

Of course, you need some way to inform the UDP layer that you want to attach this new class to a particular UDP port. You can accomplish this with the *registerProcessor* method of the *UDP* class. For example, suppose your have a new *PacketProcessor*-derived class named *Port101*. You want to attach this class to incoming port 101. You might write code like this:

```
Port101 port101=new Port101(udp);
udp.registerProcessor(port101,101);
```

Keep in mind that packets are only processed when you call *processInterface*. When a full packet is ready, the system calls your *processPacket* subroutine. However, if you don't call *processInterface*, you'll never get calls to *processPacket*.

If you need to listen to multiple ports, you can either create separate classes for each port, or you can register the same object for multiple ports. The packet contains the information you'd need to decide the action to take. The *IPPacket* (the base class for the packets) contains the following members:

■ *buffer.* The buffer that contains the packet data.

■ *destIPAddress.* The destination IP address as a four-element byte array.

■ *destPort.* The destination port.

■ *sourceIPAddress.* The source IP address as a four-element byte array.

■ *sourcePort.* The source port.

■ *packetLength.* The total length of the packet.

■ *protocol.* The protocol ID for this packet.

■ *udpLayerStart.* The start index of the UDP layer data in the buffer.

■ *getBuffer.* Returns the packet data as a byte array.

■ *getByte.* Returns a specific byte from the packet data.

■ *getWord.* Returns a specific word from the packet data.

- *maxLength.* Returns the longest possible length for the packet.
- *setByte.* Sets a specific byte in the packet data.
- *setWord.* Sets a specific word in the packet data.

This provides the information you need to dissect the packet and take whatever action required.

A Simple Sending Example

Consider a simple example where the Javelin monitors a switch and counts the number of switch closures. The count is sent to a PC at a fixed IP address on the network. You can easily build this circuit on the Javelin demo board, or you can use the schematic shown in Fig. 7-1.

The modem is a standard external modem connect via a MAX232 chip (this chip provides conversion from digital logic levels to RS232 levels). The switch simply grounds P5 on the Javelin (which has a pullup

Figure 7-1

Conecting the Javelin to a Modem

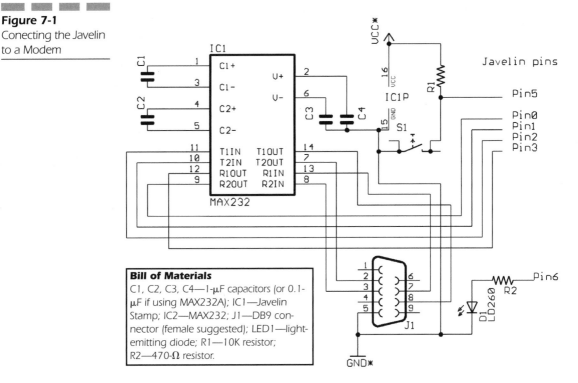

Bill of Materials
C1, C2, C3, C4—1-μF capacitors (or 0.1-μF if using MAX232A); IC1—Javelin Stamp; IC2—MAX232; J1—DB9 connector (female suggested); LED1—light-emitting diode; R1—10K resistor; R2—470-Ω resistor.

resistor attached). If you are using the Parallax demo board, you'll wire P0 to COM PORT 3, P1 to COM PORT 2, P3 to COM PORT 8, and P4 to COM PORT 7.

Listing 7-1 shows the program that monitors the switch and sends the packet. Only a few lines are required to handle the switch and the counting. The bulk of the code handles the network. Assuming you can use the standard script shown above, nearly all of the program's operation is set in the first few lines:

Listing 7-1
Monitoring a Switch

```
import stamp.core.*;
import stamp.util.*;
import stamp.util.dialer.*;
import stamp.net.*;


public class DialupSwitch {

  /**
   * Protocol stack.
   */
  private PPP ppp;
  private IP ip;
  private UDP udp;

  final static String USERNAME = "al";
  final static String PASSWORD = "password";
  final static boolean DEBUGFLAG = false;
  final static byte [] DESTIP = { 10,1,1,1};

  final static int SERIAL_RX_PIN = CPU.pin0;
  final static int SERIAL_TX_PIN = CPU.pin1;
  final static int SERIAL_CTS_PIN = CPU.pin2;
  final static int SERIAL_RTS_PIN = CPU.pin3;

  static Uart rxUart = new Uart( Uart.dirReceive,
                    SERIAL_RX_PIN, Uart.dontInvert,
                    SERIAL_CTS_PIN,Uart.speed19200,
                    Uart.stop1 );
```

Listing 7-1
Continued

```
static Uart txUart = new Uart( Uart.dirTransmit,
                        SERIAL_TX_PIN,Uart.dontInvert,
                        SERIAL_RTS_PIN,  Uart.speed19200,
                        Uart.stop1 );


static DialerAction demoScript[] = {
  new ActionPause(1000),
  new ActionPause(1000),
  new ActionSend(txUart,"+++"),        // maybe take modem offline
  new ActionWait(rxUart,"OK",3,1),
  new ActionSend(txUart, "ATZH0\r"), // hang up and reset
  new ActionWait(rxUart, "OK", 2, 1),
  new ActionPause(3000),
  new ActionSend(txUart, "ATE0V1Q0\r"),
  new ActionWait(rxUart, "OK", 2, -1),
  new ActionSend(txUart, "ATDT8888888\r"),  // ISP Phone #
  new ActionWait(rxUart, "CONNECT", 60, -6),
};


DialerControl dialer;

/**
 * Create a network stack.
 */
void initializeNetStack() {

  // Create the IP protocol.
  ip = new IP(new IPPacketPool(4));
  ppp = new PPP(ip, rxUart, txUart);
  // Set the physical interface.
  ip.iface = ppp;

  // Register the ICMP protocol (ping).
  ip.registerProcessor(new ICMP(ip), ICMP.PROTO_ICMP);
  // Register the UDP protocol.
  udp = new UDP(ip);
  ip.registerProcessor(udp, UDP.PROTO_UDP);
}
```

Listing 7-1
Continued

```
void run() {
  dialer = new DialerControl(demoScript);

  initializeNetStack();

  dialer.debug = DEBUGFLAG;
  ppp.debug=DEBUGFLAG;
  ppp.showip=true;
  if ( dialer.runScript() ) {
    System.out.println("Connection established.");

    // Now the modem is connected ask PPP to negotiate.
    ppp.open(USERNAME,PASSWORD);
    byte addr[] = DESTIP;
    Packet pkt;
    boolean up=false;
    boolean swstate=false;
    byte cnt=0;
    while ( true ) {
      ppp.processInterface();  // process network
      if (  !ppp.linkUp() )
        continue;
      if (!up) { up=true; System.out.println("Link up!"); }

// check switch status and process accordingly
      if (swstate) {
        if (CPU.readPin(CPU.pin5)) { swstate=false;
        System.out.println("Set swstate false"); }
        continue;
        }
      else
        if (!CPU.readPin(CPU.pin5)) {
          swstate=true;
          System.out.println("Set swstate true");
          cnt++;
          }
        else {
          continue;
          }
```

```
// we only get here on a switch transition
        try {
          System.out.println("Reporting!");
          pkt=udp.startPacket(addr,100,100);
          pkt.buffer[pkt.length++]=cnt;
          udp.sendPacket(pkt, true);
        }
        catch (NetException e) {
          System.out.println("Net exception!");
          ppp.close();
          break;
        }


      }
    }
  }


// This just kicks everything off
  public static void main() {
    DialupSwitch obj = new DialupSwitch();
    obj.run();
  }


}
```

```
final static String USERNAME = "al";
final static String PASSWORD = "password";
final static boolean DEBUGFLAG = false;
final static byte [] DESTIP = { 10,1,1,1};

final static int SERIAL_RX_PIN = CPU.pin0;
final static int SERIAL_TX_PIN = CPU.pin1;
final static int SERIAL_CTS_PIN = CPU.pin2;
final static int SERIAL_RTS_PIN = CPU.pin3;
```

By changing these lines, you can set the user's name and password, the
telephone number, and the target IP address (that of the PC). Don't forget that if any of the IP address numbers exceed 127, you'll need to cast
them to bytes:

```
final static byte [] DESTIP = { (byte)198,1,1,1};
```

You can also modify the debugging flag or change the pin numbers that connect to the modem.

The Javelin sends the count to port 100 of the IP address you specify. Of course, this assumes that that IP address is reachable from the computer you are dialing.

The easiest way to see this program work is to set up your own PC to act as a PPP server. Exactly how you do this varies with your version of Windows. However, for Windows XP Pro, it is relatively straightforward:

From the Start Menu, select Network Connections (be sure you are using an Administrator account). Then click on Create New Connection. This will start a Wizard (see Fig. 7-2). Click Next and pick Set Up an Advanced Connection (see Fig. 7-3) and press Next. On the screen that appears (Fig. 7-4), select Accept Incoming Connections and press Next.

The next screen will show you any devices you have available to accept connections. Select your modem from the list (see Fig. 7-5) and click Next. The next screen deals with virtual private networking (VPN). This setting won't matter for the Javelin. You can choose to allow VPNs, if you wish, and click Next (see Fig. 7-6).

Figure 7-2
Screen for New Connection Wizard

Figure 7-3

Screen for Setting Up an Advanced Connection

Figure 7-4

Screen for Accepting Incoming Connections

Figure 7-5
Screen for Choosing
Modem to Accept
'Connections

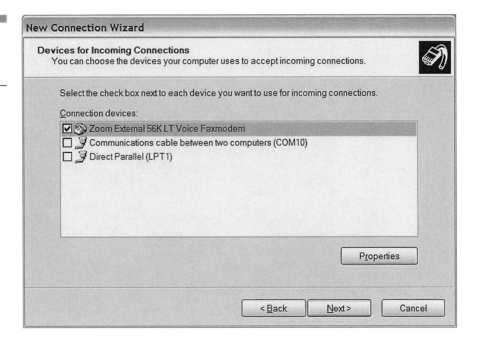

Figure 7-5
Screen for Choosing
Modem to Accept
'Connections

Figure 7-6
VPN Screen

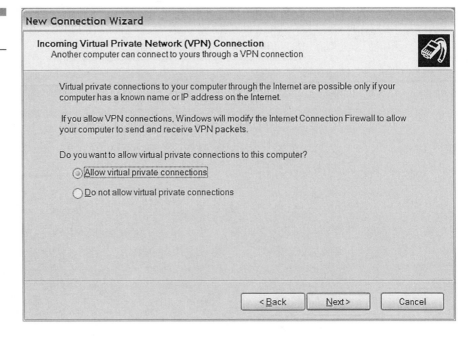

The next screen will show a list of users that are allowed to connect to your computer (see Fig. 7-7).

You can select as many users as you wish. To log in, you'll need to know the user's ID and the password for that user. Once you are done selecting users, click Next.

The next screen (Fig. 7-8) shows the different network protocols you can use. In this case, we are most interested in TCP/IP. Make sure it is checked and selected. Then press the Properties button, which will bring you to the dialog (shown in Fig. 7-9) that allows you to choose your incoming TCP/IP properties. Be sure to check "Allow callers to access my local area network." You'll probably also want to check "Assign TCP/IP addresses automatically using DHCP" unless you want specific control of the IP addresses that can be assigned to the Javelin. Dismiss this dialog by pressing OK. Then click Next and Finish.

Now your Javelin should be able to call and connect to your PC. Of course, you also need a program on the PC to read the data arriving from the Javelin. It is simple enough to write this program in normal Java (see Listing 7-2). By default, though, the Javelin won't have access to

Figure 7-7
Screen for Specifying Users Allowed to Connect to Your Computer

Figure 7-8
Screen for Choosing
Network Protocols

Figure 7-9
Screen for Choosing
Incoming TCP/IP
Properties

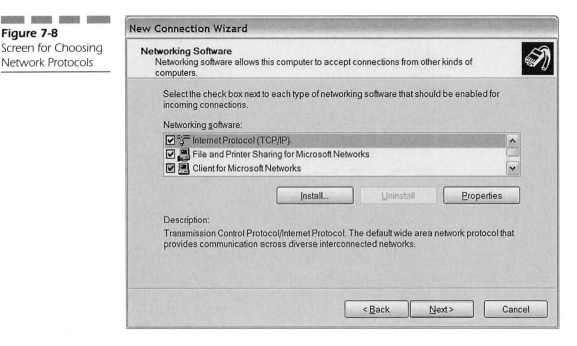

New Connection Wizard

Networking Software
Networking software allows this computer to accept connections from other kinds of computers.

Select the check box next to each type of networking software that should be enabled for incoming connections.

Networking software:

☑ 🖥 Internet Protocol (TCP/IP)
☑ 🖥 File and Printer Sharing for Microsoft Networks
☑ 🖥 Client for Microsoft Networks

[Install...]   [Uninstall]   [Properties]

Description:
Transmission Control Protocol/Internet Protocol. The default wide area network protocol that provides communication across diverse interconnected networks.

[< Back]   [Next >]   [Cancel]

Incoming TCP/IP Properties

Network access

☑ Allow callers to access my local area network

TCP/IP address assignment

◉ Assign TCP/IP addresses automatically using DHCP

○ Specify TCP/IP addresses

From:

To:

Total:

☐ Allow calling computer to specify its own IP address

[OK]   [Cancel]

Listing 7-2
Reading the Switch

```
import java.net.*;
import java.io.*;

public class PCSwitch {
    public static void main(String[] args) throws Exception {
    DatagramSocket skt=new DatagramSocket(100);
    byte [] buffer = new byte[32];
    DatagramPacket p = new DatagramPacket(buffer,32);
    while (true ) {
        skt.receive(p);
        System.out.println("From: " + p.getAddress());
            byte count = p.getData()[0];
            System.out.println("Count = " + count);
      }

    }
}
```

the public Internet, just the local network. Although you can run software on the PC that bridges to the Internet, you can also have the Javelin dial up an ISP directly.

The PC-based program simply echoes the Javelin's IP address and the count it sends. It does not reply to the Javelin, so the program is very simple. Notice that any exceptions that occur just terminate the program.

A Simple Receiving Example

Using the same basic hardware, it is simple to test the opposite proposition: making the Javelin receive commands from the network. For this demonstration you'll use the LED connected to P6. If you are using the demo board, you'll need to add an LED and the dropping resistor shown in Fig. 7-1 to the breadboard area and connect them to P6 as shown.

The idea is that a simple PC program will send the Javelin a command to turn the LED on or off. You can see the Java program in Listing 7-3. You provide the program with the Javelin's IP address and a 1 or a 0 (both on the command line). If you have the LED wired correctly, the 1 should turn on the LED and a 0 will turn it off.

Listing 7-3
Controlling the LED

```java
import java.net.*;



public class PCLEDSet {
    public static void main(String[] args) throws Exception {
DatagramSocket skt=new DatagramSocket(100);
byte [] buffer = new byte[1];
        buffer[0]=(byte)args[1].charAt(0);
DatagramPacket reply = new DatagramPacket(buffer,1,
            InetAddress.getByName(args[0]),100);
        skt.send(reply);
        skt.close();
}


    }
```

Notice that the Java program sends its data to port 100 on the Javelin. To handle this, you'll need two Java files: one for the main program, and another to provide a packet processor for port 100. The packet processor is where most of the action is, yet it is a simple class. You can see the code in Listing 7-4.

Listing 7-4
Processing Control
Messages

```java
import stamp.core.*;
import stamp.net.*;



public class LEDListener extends PacketProcessor {
  public LEDListener(PacketProcessor parent) {
    super(parent);
  }

  public void processPacket(Packet packet, int start, int length) {
    CPU.writePin(CPU.pin6,packet.getByte(start)==(byte)'1');
    System.out.println(packet.getByte(start));
  }
}
```

There are only two methods in the class. The constructor simply delegates to the superclass constructor. The other method, *processPacket,* handles the actual logic when a packet arrives. In this case, the code simply uses *writePin* to turn the LED on or off and prints a diagnostic message.

Of course, somehow you have to connect this class to port 100 in the UDP protocol. That's part of the main program's job, as seen in Listing 7-5. The actual part that connects the *LEDListener* class are these two lines:

Listing 7-5
The Main Program

```
import stamp.core.*;
import stamp.util.*;
import stamp.util.dialer.*;
import stamp.net.*;


public class LEDDemo {

  /**
   * Protocol stack.
   */
  static PPP ppp;
  static IP ip;
  static UDP udp;
  static LEDListener listener;

  final static String USERNAME = "al";
  final static String PASSWORD = "password";
  final static boolean DEBUGFLAG = false;

  final static int SERIAL_RX_PIN = CPU.pin0;
  final static int SERIAL_TX_PIN = CPU.pin1;
  final static int SERIAL_CTS_PIN = CPU.pin2;
  final static int SERIAL_RTS_PIN = CPU.pin3;

  Uart rxUart = new Uart( Uart.dirReceive, SERIAL_RX_PIN,
Uart.dontInvert,

                          SERIAL_CTS_PIN,Uart.speed19200,
                          Uart.stop1 );
```

Listing 7-5
Continued

```java
Uart txUart = new Uart( Uart.dirTransmit,
SERIAL_TX_PIN,Uart.dontInvert,
                         SERIAL_RTS_PIN,  Uart.speed19200,
                         Uart.stop1 );


DialerAction demoScript[] = {
   new ActionPause(1000),
   new ActionSend(txUart,"+++"),       // maybe take modem offline
   new ActionWait(rxUart,"OK",3,1),
   new ActionSend(txUart, "ATZH0\r"), // hang up and reset
   new ActionWait(rxUart, "OK", 2, 1),
   new ActionPause(3000),
   new ActionSend(txUart, "ATE0V1Q0\r"),
   new ActionWait(rxUart, "OK", 2, -1),
   new ActionSend(txUart, "ATDT8888888\r"),  // ISP Phone #
   new ActionWait(rxUart, "CONNECT", 60, -6),
};


DialerControl dialer;

/**
 * Create a network stack.
 */
void initializeNetStack() {

   // Create the IP protocol.
   ip = new IP(new IPPacketPool(4));
   ppp = new PPP(ip, rxUart, txUart);
   // Set the physical interface.
   ip.iface = ppp;

   // Register the ICMP protocol.
   ip.registerProcessor(new ICMP(ip), ICMP.PROTO_ICMP);

   // Register the UDP protocol.
   udp = new UDP(ip);
   ip.registerProcessor(udp, UDP.PROTO_UDP);
```

Listing 7-5

Continued

```
        listener=new LEDListener(udp);
        udp.registerProcessor(listener,100);



    }

    void run() {
      dialer = new DialerControl(demoScript);

      initializeNetStack();

      dialer.debug = DEBUGFLAG;
      ppp.debug=DEBUGFLAG;
      ppp.showip=true;
      if ( dialer.runScript() ) {
        System.out.println("Connection established.");

        // Now the modem is connected ask PPP to negotiate.
        ppp.open(USERNAME,PASSWORD);
        while ( true ) {
          ppp.processInterface(); // process network
          }
      }
    }

// Start everything
  public static void main() {
    LEDDemo demo = new LEDDemo();
    demo.run();
  }

}
```

```
listener=new LEDListener(udp);
udp.registerProcessor(listener,100);
```

Notice that the main program eventually winds up doing nothing but calling *processInterface*. That's because all the work occurs in *LEDListener* (other than the network stack setup, of course). If you had other things to do, you could perform other steps in the loop; however, in this case, there is nothing left to do. Just remember that if you don't call *processInterface,* you'll never receive any packets.

A Useful Example

For a more useful complete example, consider the code in Listing 7-6. This program uses a DS1620 temperature sensor from Maxim. Parallax provides an example class for reading this device that uses *shiftIn* and *shiftOut* to communicate with the chip.

The DS1620 is simple to wire up since it is in an eight-pin DIP package. For this project I simply used the same circuit utilized in the Inter-

Listing 7-6
Sending a Temperature

```
import stamp.core.*;
import stamp.net.*;
import stamp.util.*;
import stamp.util.dialer.*;
import stamp.peripheral.sensor.temperature.DS1620;

/**
 * Reads a temperature from a DS1620 and
 * sends it to a PC via UDP
 *
 * @version 1.0
 * @author Al Williams
 */

public class TempSender {

  // Put variables here.
  static int sent=0;  // packet #

  static PPP ppp;
  static IP ip;
  static UDP udp;
```

Listing 7-6

Continued

```
final static String USERNAME = "al";

final static String PASSWORD = "mypassword";

final static boolean DEBUGFLAG=true;

final static byte[] DESTIP = { 10,1,1,1 };

final static int destPort = 100;


final static int SERIAL_RX_PIN = CPU.pin0;

final static int SERIAL_TX_PIN = CPU.pin1;

final static int SERIAL_CTS_PIN = CPU.pin2;

final static int SERIAL_RTS_PIN = CPU.pin3;


Uart rxUart = new Uart( Uart.dirReceive, SERIAL_RX_PIN,
Uart.dontInvert,

                        SERIAL_CTS_PIN, Uart.dontInvert,
                        Uart.speed9600,

                        Uart.stop1 );
Uart txUart = new Uart( Uart.dirTransmit, SERIAL_TX_PIN,
Uart.dontInvert,

                        SERIAL_RTS_PIN, Uart.dontInvert,
                        Uart.speed9600,

                        Uart.stop1 );


DialerAction demoScript[] = {
  new ActionPause(1000),
  new ActionSend(txUart,"+++"),
  new ActionWait(rxUart,"OK",3,1),
  new ActionSend(txUart, "ATZH0\r"),
  new ActionWait(rxUart, "OK", 2, 1),
  new ActionPause(3000),
  new ActionSend(txUart, "ATE0V1Q0\r"),
  new ActionWait(rxUart, "OK", 2, -1),
  new ActionSend(txUart, "ATDT2813344341\r"),
  new ActionWait(rxUart, "CONNECT", 60, -6),
  new ActionResult(ActionResult.SUCCESS)
};


DialerControl dialer;


/**
 * Create a TCP/IP network stack.
```

Listing 7-6
Continued

```
*/
  void initializeNet() {

  // Create the IP protocol.
  ip = new IP(new IPPacketPool(4));
  ppp = new PPP(ip, rxUart, txUart);
  // Set the physical interface.
  ip.iface = ppp;

  // Register the ICMP protocol.
  ip.registerProcessor(new ICMP(ip), ICMP.PROTO_ICMP);

  // Register the UDP protocol.
  udp = new UDP(ip);
  ip.registerProcessor(udp, UDP.PROTO_UDP);
}

void run() {
  Timer clock=new Timer();
  DS1620 sensor=null;
 sensor=new DS1620(CPU.pin4,CPU.pin5,CPU.pin6);
  dialer = new DialerControl(demoScript);

  initializeNet();

  dialer.debug = DEBUGFLAG;
  ppp.debug=DEBUGFLAG;
  ppp.showip=true;
  if ( dialer.runScript() ) {
    System.out.println("Connection established.");

    // Now the modem is connected ask PPP to negotiate.
    ppp.open(USERNAME,PASSWORD);
    byte addr[] = DESTIP;
    Packet pkt;
    boolean up=false;
    int cnt=0;
    int temp;
    while ( true ) {
     try {
```

Listing 7-6
Continued

```
        ppp.processInterface();
      if (  !ppp.linkUp() )
        continue;
      if (!up) { up=true; System.out.println("Link up!"); }

      if (cnt!=0 && !clock.timeout(5000)) continue;   // only ask
every 5s
      cnt++;
      clock.mark();
      temp=sensor.getTempF();
      System.out.println("Sending: " + temp);
      pkt=udp.startPacket(addr,100,destPort);
      pkt.buffer[pkt.length++]=(byte)cnt;
      pkt.buffer[pkt.length++]=(byte)temp;


      udp.sendPacket(pkt, true);
      }
     catch (NetException e) {
       System.out.println("Net exception!");
       ppp.close();
       break;
     }


    }
   }
  }


  public static void main() {
    new TempSender().run();
  }


  static void dummy() {
    try {
      throw (NetException)NetException.throwIt();
    }
    catch (NetException e) {}
  }


}
```

net examples given earlier and also connected three pins from the DS1620 to the Javelin Stamp. In particular, pin 1 of the chip connects to the Javelin Stamp's P4; pin 2 connects to P5; and pin 3 connects to P6. Pin 4 of the DS1620 connects to ground, and pin 8 connects to 5 V. The other pins are not connected.

The sending code (see Listing 7-6) is very familiar since it follows the same template the other examples have used. The unique part simply reads the DS1620 object (named *sensor*):

```
if (sent!=0 && !clock.timeout(5000)) continue;
System.out.println("Sending");
sent++;
clock.mark();
pkt=udp.startPacket(addr,100,100);
pkt.buffer[pkt.length++]=(byte)sent;  // packet #
pkt.buffer[pkt.length++]=(byte)sensor.getTempC();  // temp
udp.sendPacket(pkt, true);  // send
```

Notice that the packet includes a sequence number. This can be used as a simple way for the client to tell if it has missed any data. It would even be possible to store samples until acknowledged so they could be retransmitted—at least, up to some limit.

The client code for the host computer (see Listing 7-7) is also straightforward. It simply echoes the value, but it would be simple enough to change the program to output to, for example, a comma delimited file (for later input to a spreadsheet program). You could also make the program store the results in a database.

Listing 7-7
Reading the Temperature

```
import java.net.*;
import java.io.*;


public class TempReader {
  static final int BUFFERSIZE = 256;
  public static void main(String [] args) {
    DatagramSocket sock;
    DatagramPacket pack=new DatagramPacket(
      new byte[BUFFERSIZE],BUFFERSIZE);
    try {
      sock=new DatagramSocket(100);
      }
```

Listing 7-7
Continued

```
catch (SocketException e) {
  System.out.println(e);
  return;
  }
while (true) {
  try {
    sock.receive(pack);
    byte packno=pack.getData()[0];
    byte temp=pack.getData()[1];
    System.out.println(packno + " - " + temp);


    }
  catch (IOException ioe) {
    System.out.println(ioe);
    }
  }
  }
}
```

If your dialup connection is to an ISP, the Javelin may well have the run of the Internet. However, if you are dialing into a PC, the Javelin may not be able to communicate with the Internet, even if the PC is connected to the Internet. One solution to this problem is to set the PC up to act as a router. However, if the PC is behind a firewall, that might not be practical since the firewall may still block some or all ports from the Javelin.

Another way to solve this difficulty is to write a custom program on the PC to bridge between the Javelin and the actual network. You could also use a custom program to send e-mail or connect to a TCP server, if you wanted to do that. Once you have the data in the PC, the possibilities are endless.

CHAPTER 8

Inside Javelin Internet

When I buy a new gadget, the first thing I usually do is open it up and look inside (unless my wife stops me, which seems to happen more often lately). After all, when it breaks down, I want to be familiar with the insides already. There was a time when personal computers were this way—you could know your system inside and out. With today's mega operating systems, high-speed motherboards, and custom chip sets, good luck! At some point you simply have to give up and trust that everything is going to work (which it sometimes does).

I think one thing that has always kept me interested in embedded systems is that you can still know the entire system from stem to stern. Granted, you might not want to plumb the depths of your real-time operating system, but you could if you wanted to do so. With that in mind, this chapter gives you a road map to the Parallax-supplied networking classes. Keep in mind that this material refers to a snapshot of the code, and the version you'll be working with will probably be slightly different. Still, the overall framework should be the same, and a look inside any version will be valuable no matter what version you are using.

The Classes

The networking library depends on several classes. There are actually just a few root classes that the other classes build upon.

Packet. The *Packet* class contains data that is either from the network or destined for the network. This class really doesn't hold any surprises. You can obtain a pointer to the packed buffer or use *getByte* and *getWord* (or the corresponding *setByte* and *setWord*) methods to manipulate the buffer indirectly.

IPPacket. This subclass of *Packet* contains some network-specific information including the source and destination addresses and ports. It also contains an offset that indicates where the start of the IP data is within the packet buffer. This is the packet type used by most of the network code. The *IPPacketPool* class manages a reusable list of *IPPacket* objects.

PacketProcessor. This class acts like a switchboard, sending packets to the correct object that should handle a particular type of message. Each receiving object is itself a *PacketProcessor*-derived object which may, in turn, dispatch the packet to other *PacketProcessor* objects.

IP. The *IP* class is a specific subclass of *PacketProcessor* for handling IP packets. It examines the protocol type in the IP header to determine which handler to call.

UDP. The *PacketProcessor* subclass examines the destination port for incoming packets and routes the packet to another *PacketProcessor* class that is registered to handle that port number.

ICMP. The ping command is the main user of the ICMP protocol. This *PacketProcessor* class does not further dispatch packets. Instead, it formulates replies to ping requests and handles them directly.

NetworkInterface. This class acts as a base class for network interfaces. It maintains the hardware address and provides placeholders for several important methods. For our purposes, the only important *NetworkInterface* is *PPP* (see below). However, Parallax is promising an Ethernet interface, and it would be possible to build other hardware layers using this base class.

PPP. This subclass of *NetworkInterface* handles the connection to a dialup ISP. It is the heavyweight class in this group and contains a great deal of code to handle the PPP state machine. You can find more details about how *PPP* works in the Appendix at the end of the book.

How It Works

The *PPP* class assumes you have already dialed your modem and entered any ISP-specific user IDs and passwords required (see Chap. 7 for more information on using the *DialerControl* object to handle this step). When you construct the *PPP* object, you tell it what *Uart* objects you want to use for transmit and receive. You also pass it a root *PacketProcessor* object.

When you are ready to open the network connection, you call *PPP.open*. If you provide a user name and password, the code will attempt to use the Password Authentication Protocol (PAP) to authenticate. Depending on your ISP, this may or may not be necessary. If it isn't required, you can simply call *open* with no arguments.

When you call *open*, the *PPP* class enters the Link Control Protocol (LCP) phase. This protocol allows both sides to negotiate regarding options. The Javelin tries to conserve as much memory and processing power as possible, so it denies most options and insists on some others. In particular, the *PPP* class allows the LCP host to require PAP authentication. When the link is established, the next phase is IP Control Protocol (IPCP). In this phase, the Javelin attempts to set a bogus IP address and expects the server to deny it (and incidentally assign the actual IP address in the process). This all occurs in the *acceptOption* method.

All of the incoming data arrives during the *processInterface* method. The main application has to call this method frequently to give the network stack a chance to handle traffic. This code assembles the PPP packets and checks that they are valid. It then sends the packets based on the protocol type. This is similar to the *PacketProcessor* logic except that the handlers for the different protocols are hard coded into the class.

When *processInterface* detects a *PROTO_IP* packet, it knows this is really an IP packet (which could be a UDP or ICMP packet). In fact, it could also be a TCP packet, but the Javelin doesn't include code to handle TCP traffic yet. The *PPP* object calls the root *PacketProcessor* (which is likely an *IP* object).

Understanding *PacketProcessor*

Network packets are like concentric rings. A PPP packet contains an IP packet. The IP packet might contain a UDP packet. The *PacketProcessor* class works the same way. One processor object can be the parent of another object. When the hardware layer calls the root processor, it will examine some part of the packet and find another *PacketProcessor* that wants to handle that type of packet. To do this, the object's *dispatchPacket* routine compares the *protocol* argument (this depends on what type of packet is being used; for example, it might be a UDP port number) to a list of registered objects that want to receive packets. The receiving *PacketProcessor* may identify another protocol and then search its list of processors, passing the packet further up the chain until finally a processor decides to handle the packet.

Armed with this information, it is easier to understand the *initializeNet* method from Chap. 7:

```
void initializeNet() {

// Create the IP protocol.
ip = new IP(new IPPacketPool(4));
ppp = new PPP(ip, rxUart, txUart);
```

```
        // Set the physical interface.
        ip.iface = ppp;

        // Register the ICMP protocol.
        ip.registerProcessor(new ICMP(ip), ICMP.PROTO_ICMP);
        // Register the UDP protocol.
        udp = new UDP(ip);
        ip.registerProcessor(udp, UDP.PROTO_UDP);
    }
```

Here, the program creates an *IP* packet processor that has room to hold four packets. As processing completes, the code reuses these packets. Next, the program creates a *PPP* object, using two *Uart* objects and the *IP* object created in the first step. Finally, the program registers two more packet processors (an *ICMP* and a *UDP* processor) with the *IP* object. When *IP* packets arrive, the *PPP* object passes them to the *IP* layer. The *IP* layer then sends each packet to one of the other processors, depending on its type.

Of course, for your program to handle UDP packets (as the examples in Chap. 7 did), you'll have to create your own *PacketProcessor* subclass. Then you register your custom class with the *UDP* processor and the port number you wish to handle.

Highlights from *PPP*

Listing 8-1 shows the Parallax-supplied *PPP* class as it existed at the time of this writing, with some additional comments (marked with ✶✶✶). You can use it as a reference as you read through this chapter, although be sure to use the latest version when actually compiling your programs.

Listing 8-1
The PPP Class

```
// *** Look for special comments marked this way
package stamp.net;


import stamp.core.*;


public class PPP extends NetworkInterface {


// *** Each PPP packet starts with a flag (and typically
// ends with a flag too, although the ending flag can also
```

```
// optionally start the next packet).
  final static int PPP_FLAG = 0x7E;
// *** Characters in a certain range are escaped by
// using this value and then being XOR'd with 0x20
  final static int PPP_ESCAPE = 0x7D;
  final static int PPP_XOR = 0x20;
  final static int PPP_ADDRESS = 0xFF;
  final static int PPP_CONTROL = 0x03;

  // negPhase
  // *** Phases: Link control, IP control, IP, and
  // Password authentication
  final static int PROTO_LCP = (short)0xC021;
  final static int PROTO_IPCP = (short)0x8021;
  final static int PROTO_IP = (short)0x0021;
  final static int PROTO_PAP = (short)0xc023;

  // pppState
  // *** State machine states
  final static int PPP_STATE_INITIAL = 0;
  final static int PPP_STATE_STARTING = 1;
  final static int PPP_STATE_CLOSED = 2;
  final static int PPP_STATE_STOPPED = 3;
  final static int PPP_STATE_CLOSING = 4;
  final static int PPP_STATE_STOPPING = 5;
  final static int PPP_STATE_REQ_SENT = 6;
  final static int PPP_STATE_ACK_RCVD = 7;
  final static int PPP_STATE_ACK_SENT = 8;
  final static int PPP_STATE_OPENED = 9;

  final static int PPP_CONF_REQ = 1;
  final static int PPP_CONF_ACK = 2;
  final static int PPP_CONF_NAK = 3;
  final static int PPP_CONF_REJ = 4;
  final static int PPP_TERM_REQ = 5;
  final static int PPP_TERM_ACK = 6;
  final static int PPP_CODE_REJ = 7;
  final static int PPP_PROTO_REJ = 8;
```

Listing 8-1
Continued

```
final static int PPP_ECHO_REQ = 9;
final static int PPP_ECHO_REPLY = 10;
final static int PPP_DISCARD_REQ = 11;

final static int PPP_PROTO_FIELD = 2;
final static int PPP_INFO_FIELD = 4;
final static int PPP_CODE_FIELD = 4;
final static int PPP_ID_FIELD = 5;
final static int PPP_LENGTH_FIELD = 6;
final static int PPP_OPTION_TYPE_FIELD = 8;
final static int PPP_OPTION_LENGTH_FIELD = 9;
final static int PPP_IP_FIELD = 10;

final static int PPP_EVENT_NONE = 0;
final static int PPP_EVENT_TO_GOOD = 1;
final static int PPP_EVENT_TO_BAD = 2;
final static int PPP_EVENT_RCR_GOOD = 3;
final static int PPP_EVENT_RCR_BAD = 4;
final static int PPP_EVENT_RCA = 5;
final static int PPP_EVENT_RCN = 6;
final static int PPP_EVENT_RTR = 7;
final static int PPP_EVENT_RTA = 8;
final static int PPP_EVENT_RUC = 9;
final static int PPP_EVENT_RXJ_GOOD = 10;

final static int LCP_OPTION_MRU = 1;
final static int LCP_OPTION_AUTHENICATION = 3;

final static int IPCP_OPTION_ADDR = 3;

boolean escaped = false;

// *** This sets the maximum size packet
public final static int MAX_PACKET_SIZE = 254;
// *** Set this to true to log debug messages
public static boolean debug = false;
// *** Set this to log the assigned IP address
public static boolean showip = false;
```

Listing 8-1
Continued

```
Uart rxUart, txUart; //*** UARTs

// The packet currently being received.
Packet packet;
int pos;

// The number of packets dropped due to no free buffers.
int droppedPackets = 0;

// The FCS of the packet under construction.
int fcs;

// The negotiation phase (LCP, PCP, IPCP).
int negPhase;
// PPP state machine state.
int pppState;
// The ID of the last packet sent.
int lastID;
// Whether to do PAP
boolean doPAP;

Timer timer = new Timer();  // *** Track timeout
int retryCount;

public static byte ipAddr[] = new byte[4];
IP ip; // Reference to the IP layer.

public static PPP iface;

public PPP(IP ip, Uart rxUart, Uart txUart) {
  this.ip = ip;
  this.rxUart = rxUart;
  this.txUart = txUart;
  iface = this;
}

boolean checkCRC( Packet p, int length ) {
  int checksum = (short)0xffff;
```

Listing 8-1
Continued

```
    for ( int i = 0; i < length; i++ ) {
        checksum = addCRC(checksum, p.getByte(i));
    }
    return checksum == (short)0xF0B8;
}


public boolean linkUp() {
    return negPhase == PROTO_IP;
}


int addCRC(int crc, int b) {
    int i;
    b&=0xFF;  // adjust for signed byte
    b ^= crc;
    b &= 0xff;
    for (i = 0; i < 8; i++)
        b = ((b&1)!=0)?((b>>>1)^(short)0x8408):(b>>>1);


    return (crc>>>8) ^ b;
}


/**
 * Process any packets received on the PPP interface.
 *
 * @returns false if the link goes down. true otherwise.
 */
public boolean processInterface() {
    int event = PPP_EVENT_NONE;

    while ( rxUart.byteAvailable() ) {
        byte b = (byte)rxUart.receiveByte();

        if ( b == PPP_FLAG ) {
            if ( packet != null) {

                if ( pos == 0 )
                    continue;
```

Listing 8-1
Continued

```
          try {
              // Found the end of a packet.
              if ( !checkCRC(packet,pos)) {
                  // Bad FCS.
                  if (debug) System.out.println("Bad FCS");
// Discard packet.
                  throw RuntimeException.throwIt();
              }


              // Compute the event based on the packet.
              int proto = packet.getWord(PPP_PROTO_FIELD);
              int code = packet.getByte(PPP_CODE_FIELD);


              if ( proto == PROTO_IP ) {
                  ip.processPacket(packet,PPP_INFO_FIELD,
                      pos-6);
// Discard silently.

                  throw RuntimeException.throwIt();
}

              if ( proto != negPhase )
// Discard silently.
                  throw RuntimeException.throwIt();
              if ( code == PPP_CONF_REQ ) {
                  if (debug)
                    System.out.println("Generating reply");
                  event = generateReply( packet );
              }


              if ( code == PPP_CONF_ACK )
                  event = PPP_EVENT_RCA;


              if ( code == PPP_CONF_NAK) {
                  event = PPP_EVENT_RCN;
                  if ( negPhase == PROTO_IPCP ) {
                      for ( int i = 0; i < 4; i++ )
                          ipAddr[i] =
```

Listing 8-1
Continued

```
                          (byte)packet.getByte(PPP_IP_FIELD+i);
              if (debug || showip) {
                System.out.print("IP Addr: ");
                System.out.print(((int)ipAddr[0])&0xFF);
                System.out.print(".");
                System.out.print(((int)ipAddr[1])&0xFF);
                System.out.print(".");
                System.out.print(((int)ipAddr[2])&0xFF);
                System.out.print(".");
                System.out.println(((int)ipAddr[3])&0xFF);
              }
              ip.setIPAddress(ipAddr);
              for ( int i = 0; i < 4; i++ )
                 optionIPCP[2+i] =

                 packet.getByte(PPP_IP_FIELD+i);
          }
        }


      if ( code == PPP_CONF_REJ ) {
         event = PPP_EVENT_RXJ_GOOD;
         if ( negPhase == PROTO_LCP )
            optionLCP = optionNone;
         else if ( negPhase == PROTO_IPCP )
            optionIPCP = optionNone;
         sendRequest();
      }


      processEvent(event);
    }
    catch (RuntimeException e) { } // Do nothing.
    finally {
      ip.packetPool.checkIn(packet);
      packet = null;
    }
  }
  // Start a new packet.
  pos = 0;
```

Listing 8-1
Continued

```
    try {
        packet = (Packet)ip.packetPool.checkOut();
    }
    catch (IndexOutOfBoundsException e) {
// If there are no spare buffers, drop the packet.
        droppedPackets++;
        if (debug)
            System.out.println(
                "No buffers, dropped packet");
    }

}
else {
    if ( escaped ) {
        b ^= PPP_XOR;
        escaped = false;
    }
    else if ( b == PPP_ESCAPE ) {
        escaped = true;
        break;
    }
    else if ( b>=0 && b < PPP_XOR ) {
        // Drop any characters < 0x20.
        continue;
    }

    if ( packet == null )
        continue;

    packet.setByte(pos++, b);
    if ( pos >= packet.maxLength() ) {
        // The packet is too long. Drop it.
        droppedPackets++;
        ip.packetPool.checkIn(packet);
        packet = null;
        if (debug)
            System.out.println("Dropped oversized packet");
    }
```

Listing 8-1
Continued

```java
      }
    }
// *** If IP hasn't been established in the timeout period
// something is wrong so retry (up to 4 times)
    if ( negPhase != PROTO_IP && timer.timeout(20,0) ) {
      if ( retryCount++ >= 4 ) {
        if (debug) System.out.println("PPP aborting");
        return false;
      }
      event = PPP_EVENT_TO_GOOD;
      timer.mark();
      if (debug) System.out.println("PPP retry");
      processEvent(event);
    }

    return true;
  }

  /**
   * Determine whether an option is acceptable.
   *
   * We accept only two options:
   * 1. Authentication (PAP) during LCP.
   * 2. IP address assignment during IPCP.
   *
   * @returns one of PPP_CONF_ACK, PPP_CONF_NAK or PPP_CONF_REJ
   */
  int acceptOption(int option, Packet p, int start) {
    if ( negPhase == PROTO_LCP
        && ( option == LCP_OPTION_AUTHENICATION ) ) {
      if ( p.getWord(start) == PROTO_PAP ) {
        doPAP = true;
        if (debug) System.out.println("Found PAP Request");
        return PPP_CONF_ACK;
      }
      else {
        if (debug)
          System.out.println("Unknown Auth Request");
```

Listing 8-1
Continued

```
       return PPP_CONF_NAK;
     }
   }
   else if ( negPhase == PROTO_IPCP
           && ( option == IPCP_OPTION_ADDR ) )
     return PPP_CONF_ACK;
   else
     return PPP_CONF_REJ;
}


/**
 * Given a configure request, determine the appropriate
 * reply and event.
 */
int generateReply(Packet p) {
  int length = packet.getWord(PPP_LENGTH_FIELD);
  int action = PPP_CONF_ACK, newAction;

  // Loop through the packet checking each option.
  for ( int i = PPP_OPTION_TYPE_FIELD; i <
    PPP_INFO_FIELD+length; ) {
    int optionType = packet.getByte(i);
    int optionLength = packet.getByte(i+1);
    newAction = acceptOption(optionType, packet, i+2);
    if ( newAction == PPP_CONF_NAK &&
         action == PPP_CONF_ACK )
      action = PPP_CONF_NAK;
    else if ( newAction == PPP_CONF_REJ )
      action = PPP_CONF_REJ;

    i += optionLength;
  }

  // Now, depending on the accepted options form the
  // reply packet.
  // Any option that matches the reply type (ACK, NAK or
  // REJ) is copied
  // to the reply packet.
```

Listing 8-1
Continued

```
    int bPos = PPP_OPTION_TYPE_FIELD;
    for ( int i = PPP_OPTION_TYPE_FIELD;
          i < PPP_INFO_FIELD+length; ) {
      int optionType = packet.getByte(i);
      int optionLength = packet.getByte(i+1);


      newAction = acceptOption(optionType, packet, i+2);
      // The option matches the packet action. Copy it.
      if ( newAction == action ) {
        if (action==PPP_CONF_NAK)
          packet.setWord(i+2,PROTO_PAP);
        for ( int j = i; j < i + optionLength;
              bPos++, j++ )
          packet.setByte(bPos, packet.getByte(j));
        }
      i += optionLength;
    }
    packet.setByte(PPP_CODE_FIELD, (byte)action);
    packet.setWord(PPP_LENGTH_FIELD, bPos-PPP_INFO_FIELD);
    p.length = bPos;
      if (debug) {
        switch (action) {
          case PPP_CONF_ACK:
            System.out.println("Sent PPP_CONF_ACK"); break;
          case PPP_CONF_NAK:
            System.out.println("Sent PPP_CONF_NAK"); break;
          case PPP_CONF_REJ:
            System.out.println("Sent PPP_CONF_REJ"); break;
          }
        }

    if ( action == PPP_CONF_ACK )
      return PPP_EVENT_RCR_GOOD;
    else
      return PPP_EVENT_RCR_BAD;
  }

  void sendRequest() {
```

Listing 8-1
Continued

```
  switch ( negPhase ) {
  case PROTO_LCP:
    sendPacket(PROTO_LCP, PPP_CONF_REQ,
      ++lastID, optionLCP.length, 0, optionLCP);
    if (debug) System.out.println("Sent optionLCP");
    break;

  case PROTO_IPCP:
    sendPacket(PROTO_IPCP, PPP_CONF_REQ, ++lastID,
      optionIPCP.length, 0, optionIPCP);
    if (debug) System.out.println("Sent optionIPCP");
    break;

  case PROTO_PAP:
    sendPacket(PROTO_PAP, PPP_CONF_REQ, ++lastID,
      optionPAP.length, 0, optionPAP);
    if (debug) System.out.println("Sent optionPAP");
    break;
  }
}

void processEvent(int event) {

  switch (pppState) {
  case PPP_STATE_REQ_SENT:
    switch (event) {
    case PPP_EVENT_RCR_BAD:
      sendReplyPacket( packet );
      break;

    case PPP_EVENT_RCR_GOOD:
      sendReplyPacket( packet );
      pppState = PPP_STATE_ACK_SENT;
      retryCount = 0;
      break;
    case PPP_EVENT_RCA:
      pppState = PPP_STATE_ACK_RCVD;
      retryCount = 0;
```

Listing 8-1

Continued

```
      break;
   case PPP_EVENT_RCN:
     if ( negPhase == PROTO_IPCP )
       sendRequest();
     else if ( negPhase == PROTO_LCP ) {
       optionLCP = optionNone;
       sendRequest();
     }
     break;
   case PPP_EVENT_TO_GOOD:
     sendRequest();
     break;
   case PPP_EVENT_TO_BAD:
     break;
   }
   break;
 case PPP_STATE_ACK_RCVD:
   switch (event) {
   case PPP_EVENT_RCR_GOOD:
     sendReplyPacket( packet );
     if ( negPhase == PROTO_LCP && doPAP ) {
       negPhase = PROTO_PAP;
       sendRequest();
       pppState = PPP_STATE_ACK_SENT;
     }
     else if ( negPhase == PROTO_LCP ||
               negPhase == PROTO_PAP ) {
       negPhase = PROTO_IPCP;
        sendRequest();
       // Since authentication is one sided
       // (we authenticate with the server,
       // the server doesn't authenticate with us) we
       // need to fool the state
       // machine into thinking that half of the
       // negotiation is already done.
       pppState = PPP_STATE_REQ_SENT;
     }
     else if ( negPhase == PROTO_IPCP ) {
```

Listing 8-1
Continued

```
          negPhase = PROTO_IP;
          if (debug) System.out.println("PPP Link Up");
      }
    break;
  case PPP_EVENT_RCA:
    pppState = PPP_STATE_REQ_SENT;
    retryCount = 0;
  case PPP_EVENT_RCR_BAD:
    sendReplyPacket( packet );
    break;
  case PPP_EVENT_RCN:
    if ( negPhase == PROTO_IPCP ) {
        sendRequest();
    }
    else if ( negPhase == PROTO_LCP ) {
      optionLCP = optionNone;
      sendRequest();
    }
    pppState = PPP_STATE_REQ_SENT;
    retryCount = 0;
    break;
  }
    break;
case PPP_STATE_ACK_SENT:
    switch (event) {
    case PPP_EVENT_RCR_BAD:
      sendReplyPacket( packet );
      pppState = PPP_STATE_REQ_SENT;
      retryCount = 0;
      break;
    case PPP_EVENT_RCR_GOOD:
      sendReplyPacket( packet );
      break;
    case PPP_EVENT_RCA:
      if ( negPhase == PROTO_LCP && doPAP ) {
        negPhase = PROTO_PAP;
        sendRequest();
        pppState = PPP_STATE_ACK_SENT;
```

Listing 8-1
Continued

```
            }
            else if ( negPhase == PROTO_LCP ||
                        negPhase == PROTO_PAP ) {
                negPhase = PROTO_IPCP;
                sendRequest();
                pppState = PPP_STATE_REQ_SENT;
            }
            else if ( negPhase == PROTO_IPCP ) {
                negPhase = PROTO_IP;
                if (debug) System.out.println("PPP Link Up");
            }
            break;
        case PPP_EVENT_RCN:
            if ( negPhase == PROTO_IPCP )
                sendRequest();
            else if ( negPhase == PROTO_LCP ) {
                optionLCP = optionNone;
                sendRequest();
            }
            break;
        case PPP_EVENT_TO_GOOD:
            sendRequest();
            break;
        }
    }
}


// *** options for the different phases (see text)
    static int optionNone[] = {};
    static int optionLCP[] =
        {LCP_OPTION_MRU,4,0,MAX_PACKET_SIZE};
    static int optionPAP[] = {1,'x',1,'y',};
    static int optionIPCP[] = {IPCP_OPTION_ADDR,6,0,0,0,0};

    /**
     * Initiate a PPP connection.
     */
// *** This version opens
```

Listing 8-1

Continued

```
public void open(String uid, String pw) {
optionPAP = new int[uid.length()+pw.length()+2];
optionPAP[0]=uid.length();
int i,j;
for (i=0;i<uid.length();i++)
   optionPAP[i+1]=uid.charAt(i);
optionPAP[i+1]=pw.length();
for (j=0;j<pw.length();j++)
   optionPAP[i+j+2]=pw.charAt(j);
if (debug) System.out.println("Using uid=" + uid +
   " pw=" + pw);
open();
}


// *** This kicks off the state machine
// and starts timing for timeout
  public void open() {
    lastID = 0;
    negPhase = PROTO_LCP;
    sendRequest();
    pppState = PPP_STATE_REQ_SENT;
    timer.mark();
    retryCount = 0;
    doPAP = false;
  }


  public void close() {
// *** Tell peer we are closing
    negPhase = PROTO_LCP;
    sendPacket(PROTO_LCP, PPP_TERM_REQ, ++lastID,
               0, 0, null );
  }


  void sendPacket(int protocol, int code, int id,
    int length, int offset, int data[]) {
    fcs = (short)0xffff;  // *** Start FCS
    // Send the PPP header.
// *** sendByte will update FCS
```

Listing 8-1
Continued

```
    txUart.sendByte(PPP_FLAG);
    sendByte(PPP_ADDRESS);
    sendByte(PPP_CONTROL);
    sendByte(protocol>>>8);m
    sendByte(protocol&0x00ff);

    // Send the packet data.
    sendByte(code);
    sendByte(id);
    int len = length + 4;
    sendByte(len>>>8);
    sendByte(len&0x00ff);
    for ( int i = 0; i < length; i++ )
        sendByte(data[i+offset]);

    // Send the FCS.
    int temp = ~fcs;
    sendByte(temp&0x00ff);
    sendByte(temp>>>8);
    txUart.sendByte(PPP_FLAG);
}

public void sendPacket(Packet p, int start) {
    fcs = (short)0xffff;
    // Send the PPP header.
    txUart.sendByte(PPP_FLAG);
    sendByte(PPP_ADDRESS);
    sendByte(PPP_CONTROL);
    sendByte(PROTO_IP>>>8);
    sendByte(PROTO_IP&0x00ff);

    for ( int i = start; i < p.length + start; i++ )
        sendByte(p.getByte(i));

    // Send the FCS.
    int temp = ~fcs;
    sendByte(temp&0x00ff);
    sendByte(temp>>>8);
```

262

Chapter 8

Listing 8-1
Continued

```
        txUart.sendByte(PPP_FLAG);
    }

    public void sendReplyPacket(Packet p) {
        fcs = (short)0xffff;
        // Send the PPP header.
        txUart.sendByte(PPP_FLAG);

        for ( int i = 0; i < p.length; i++ )
            sendByte(p.getByte(i));

        // Send the FCS.
        int temp = ~fcs;
        sendByte(temp&0x00ff);
        sendByte(temp>>>8);
        txUart.sendByte(PPP_FLAG);
    }

    void sendByte(int b) {
        fcs = addCRC(fcs, b);
        if ((b>=0 && b < 0x20) || b == PPP_FLAG ||
            b == PPP_ESCAPE ) {
            txUart.sendByte(PPP_ESCAPE);
            b ^= 0x20;
        }
        txUart.sendByte(b);
    }

}
```

The *PPP* class has several important methods:

- *PPP.* The constructor requires a root *PacketProcessor* instance, and two *Uart* objects (one for transmit and one for receive).
- *checkCRC.* This routine verifies the cyclic redundancy check (CRC) of a packet. This detects if there were errors in the transmission.
- *linkUp.* Your program can call *linkUp* to determine if the network is ready to use.

- *addCRC.* This routine actually accepts a byte and adds it to a CRC to produce a new CRC value. Both sending and receiving routines use this to compute CRCs.

- *processInterface.* The main program calls this method to allow the network stack to process data. It actually receives data from the UART, assembles the data into PPP packets, and either processes them or passes them to the IP packet processor.

- *acceptOption.* This method examines configuration options and decides if it should accept them (*PPP_CONF_ACK*), reject them (*PPP_CONF_REJ*), or deny even knowing about them at all (*PPP_CONF_NAK*).

- *generateReply.* When a string of options arrives, this routine splits them up into separate options and calls *acceptOption.* If any options are *PPP_CONF_REJ,* that becomes the return code for the entire reply. Likewise, if there are no rejections, but any option generates *PPP_CONF_NAK,* that becomes the return code. Once the method determines the return code, it sweeps through the list again, generating a new list that comprises the options that the return code indicates.

    Suppose there are four options sent (call them A, B, C, and D). If options C and D generate a *PPP_CONF_REJ,* this routine will formulate a response that indicates rejection with a list containing C and D only. This indicates to the sender which options you won't accept.

    The Javelin is also responsible for telling the remote side what options it wants or what values it would find acceptable. You can find the defaults in the *static* variables *optionLCP, optionPAP,* and *optionIPCP.* In particular, the code requests a maximum receive unit (MRU) of *MAX_PACKET_SIZE,* and an IP address of 0.0.0.0 (which will be rejected, of course). The *optionPAP* variable contains a default user ID and password, but the *open* method rewrites this to contain your user ID and password.

- *sendRequest.* This routine sends packets to respond to LCP, IPCP, or PAP messages.

- *processEvent.* During *processInterface,* several conditions constitute events. These events correspond to transition events in the PPP standard (see Appendix).

- *open.* There are two *open* calls. One takes a user ID and password. The other requires no arguments and assumes you've already authenticated with your ISP in some way. This method kicks off the PPP state machine. You still need to call *processInterface* before anything really happens. You should always call *linkUp* and receive a *true* response before trying to use the connection.

- *close*. This method closes the connection and sends a packet to inform the other computer that the stack is closing down.

- *sendPacket*. There are two versions of *sendPacket*, although they both do what you'd expect: send a packet to the remote computer. One version simply takes a *Packet* (assumed to be an IP packet) and a starting offset. The other takes raw data and builds the packet (for any protocol) for you before sending it. Both routines also send the required PPP header, the CRC code (known as the frame check sequence or FCS), and the trailing PPP flag byte.

- *sendReplyPacket*. When you receive a packet, it has much of the required data already in place, so this method allows you to simply make changes in place and send the packet back.

- *Send Byte*. Each of the methods that sends a packet actually uses *sendByte* to transmit the byte to the remote computer. The method escapes special characters using the *PPP_ESCAPE* character and computes the CRC.

Inside UDP

The *UDP* class allows you to register a class that will receive packets destined for a specific UPD port. Remember, this class is derived from *Packet-Processor* and expects your custom classes to also be *PacketProcessor*-derived.

You'll find the version of the UDP class that was current at the time of this writing in Listing 8-2. You can study this listing along with the overview below it, but be sure to use the most current version when compiling your programs. You'll find additional comments marked with three asterisks.

There are several important methods in the *UDP* class:

- *UDP*. The constructor requires a reference to the *PacketProcessor* that will send the *UDP* object packets.

- *processPacket*. This method receives packets and determines the source and destination ports. The method calls *dispatchPacket* (provided by the base class) which will, in turn, pass the packet to another *processPacket* in your custom class, if you have a class registered for the destination port.

```
// *** Look for additional comments marked this way
package stamp.net;

public class UDP extends PacketProcessor {

  // *** The protocol ID for UDP is 17
  public final static int PROTO_UDP = 17;

// *** Offsets within the packet
// for the source port, destination port, etc.
  final static int SOURCE_PORT_FIELD = 0;
  final static int DEST_PORT_FIELD = 2;
  final static int LENGTH_FIELD = 4;
  final static int CHECKSUM_FIELD = 6;
  final static int DATA_FIELD = 8;


  final static int UDP_HEADER_LENGTH = 8;

  public UDP(PacketProcessor parent) {
    super(parent);
  }

// *** Look at the destination port and
// use it to find a child processor for this packet
// dispatchPacket is in the base class
  public void processPacket(Packet packet, int start,
        int length) {
    IPPacket p = (IPPacket)packet;

    p.destPort = p.getWord(start+DEST_PORT_FIELD);
    p.sourcePort = p.getWord(start+SOURCE_PORT_FIELD);

    dispatchPacket(p, p.destPort,
        start+UDP_HEADER_LENGTH, p.length);
  }

  public Packet startReply(Packet packet)
    throws NetException {
```

Listing 8-2

Continued

```
IPPacket reply =
    (IPPacket)parentProcessor.startReply(packet);
IPPacket p = (IPPacket)packet;
int len = reply.length;

reply.udpLayerStart = len;
reply.setWord(len+SOURCE_PORT_FIELD, p.destPort);
reply.setWord(len+DEST_PORT_FIELD, p.sourcePort);
reply.setWord(len+LENGTH_FIELD, UDP_HEADER_LENGTH);

reply.setWord(len+CHECKSUM_FIELD,0);

reply.length += UDP_HEADER_LENGTH;

return reply;
}

public Packet startPacket( byte destAddress[],
  int sourcePort, int destPort)
  throws NetException {
  IPPacket reply =
    (IPPacket)parentProcessor.startPacket(
    PROTO_UDP, destAddress);
  int len = reply.length;

  reply.udpLayerStart = len;
  reply.setWord(len+SOURCE_PORT_FIELD, sourcePort);
  reply.setWord(len+DEST_PORT_FIELD, destPort);

  reply.setWord(len+CHECKSUM_FIELD,0);

  reply.length += UDP_HEADER_LENGTH;

  return reply;
}

public void sendPacket(Packet packet, boolean free)
    throws NetException {
```

Listing 8-2
Continued

```
        IPPacket p = (IPPacket)packet;


        p.setWord(p.udpLayerStart+LENGTH_FIELD,
        p.length-p.udpLayerStart);
        if ( parentProcessor != null )
          parentProcessor.sendPacket(packet, free);
      }


    }
```

■ *startReply.* The *startReply* method starts a packet that is addressed to another computer that has already sent your program a packet. Using this method prevents you from having to explicitly know the address and port number of the sender.

■ *startPacket.* To originate a packet, you can use this method with an IP address, a source port, and a destination port number.

■ *sendPacket.* Once you've filled in the rest of a packet, you can call *sendPacket* to actually transmit the data to the remote computer.

What's Next?

There are several items the Javelin's networking classes don't address—at least not yet. Some of these may already be in the works and available by the time you read this. The most notable absence is TCP networking. You could derive a new *PacketProcessor* for TCP, register it with the *IP* object and handle TCP. However, this would not be a task for the faint of heart. Handling TCP requires management of sockets and a complex state machine that would make this a formidable task.

In addition, the Javelin doesn't handle IP fragmentation. This could pose a problem on some networks, especially those that might use a transmission size even smaller than the Javelin's. This wouldn't be very hard to add to the *IP* object, but it would be memory-intensive since incomplete packets would require storage.

Another piece missing is a connection to other types of devices and dialup protocols. So if you need to interface with Ethernet or even Serial

Line IP (SLIP), there is no code for that. Granted, the framework is there (and Parallax will probably have an Ethernet solution by the time you read this), but the baseline code lacks these features.

Still, for embedded systems connected to the telephone, PPP and UDP are very useful. If you really need to send e-mail or serve Web pages, you can do that with a remote computer, hosted anywhere in the world, that receives UDP packages and postprocesses them (an approach you'll see used by another vendor in Chap. 9).

Ready-Made Internet Appliances

Some people build beautiful furniture from raw wood. Not only are they skilled craftspeople, but they also have plenty of extra time! Most of us limit our furniture building to the furniture you take home from your local Wal-Mart or K-Mart in a box and assemble from the plans provided. Of course, what you would prefer is to buy high-quality furniture and have it delivered in a big truck ready to use.

Networking stacks are somewhat like furniture. It is possible to roll a stack from scratch, but it is highly specialized work and requires a great deal of time to get it all working properly. Solutions like the Javelin or even the TINI are more similar to Wal-Mart kits. They give you the basic components and you put them together to get what you want.

However, there is another option—buying a complete TCP/IP stack in a piece of hardware. These stacks are not meant to be processors (like the TINI) and they don't require you to consume a lot of code and data space to get connectivity. Instead, you simply hook up your processor to these prebuilt components and get on the Internet.

There are many devices like this available, and the list is growing every day. If your time is valuable to you, these may offer the easiest way to Internet-enable a microprocessor. This chapter will show you a specialized modem that knows how to send e-mail (the iModem from Cermetek).

The iModem

At first glance, the iModem seems a little simplistic. The modem is a black box [not very large, but larger than an integrated circuit (IC)] with some pins sticking out the bottom. It looks like an oversized IC with 22 pins, and that's how you can think of it. Inside the box is a Hayes-compatible modem, all the circuitry required to connect to a phone line, and enough smarts to send and receive e-mail.

Of course, you might want to do something other than send e-mail. Cermetek makes a service available to you (although not for free) that allows you to send specially formatted e-mail to their servers and have it translated to a Web page, transmitted via FTP, and sent as a FAX or voice mail. Of course, the system can also forward the e-mail to other recipients.

On the receiving end, the iModem can receive e-mail from a POP3 mailbox (which can also be supplied by the Cermetek service known as Press4Service). As you might expect, there are limits to how large a message you can receive (or send, for that matter). However, for many purposes the iModem fills the bill nicely.

Sending e-mail with the iModem is straightforward. However, receiving it seems a little rough—at least with the current version of the iModem. Still, it does work if you work with the modem carefully.

Challenges

There are a few items you need to account for when using the iModem:

- The modem requires you to use a certain baud rate (either 2400 baud or 57,600 baud, depending on the model).

- You must provide a lengthy delay (50 ms) between each character to allow the modem to process the character. You also must delay about 200 ms between each command (or more for commands that dial up, of course).

- The modem requires about 10 s after powerup before it will accept commands.

- The messages the modem returns are text strings, which makes it difficult to process in a microcontroller.

- Each operation requires the modem to dial the ISP and hang up. You can't, for example, make one phone call, send two messages, retrieve all pending messages, and delete them. If there were three pending messages, you'd need at least eight phone calls to perform those operations.

- The modem (or the ISP) doesn't correctly disconnect POP3 (e-mail reception) transactions. This means that successive e-mail retrieval (or deletion) operations may fail. Cermetek told me that the POP3 server may require up to 15 min to reset, although in informal testing it usually took 5 to 10 min.

- The modem can't use hardware handshaking if it is in Internet mode.

Some of these problems are more serious than others. Part of the problem is the modem almost seems designed for a human user. Matching the specified baud rate is no problem, and it is fairly simple to pace sent characters and parse through the text messages. Unfortunately, the repeated dialing and failed operations are a bit harder to deal with. Each dialup means a new connection sequence, so failed connects will eventually disturb your plans. In addition, each time the POP3 server locks the modem out, you have to wait for things to reset and there's nothing you can do about it.

Setup

The modem requires setup to tell it the ISP's phone number, the user ID and password for the mail servers, and other infrequently changed data. Cermetek provides a PC-based program to set these parameters, or you can use a normal terminal program (like Windows Hyperterminal). Of course, you could make the embedded system set everything, but since these items rarely change, you can usually set them once and forget about them.

All Internet-related modem commands start with *@T*. The modem also responds to the standard *AT* command that most modems do (you can also use the iModem as a regular modem). Here are some of the most common setup commands:

@TA1.	Enter destination e-mail address.
@TE1.	Enter modem's e-mail address.
@TJ0.	Set default message when SEND pin asserted (see below).
@TJ1.	Set custom message when SEND pin asserted (see below).
@TL1.	Set ISP user name.
@TN1.	Set ISP phone number.
@TP1.	Set ISP password.
@TOP1.	Set POP3 server address (IP address in hex).
@TOS1.	Enter SMTP server address (IP address in hex).

Most of the configuration parameters also accept a query format, so to set the subject line you can issue this command:

```
@TS1=Auto Message
```

If you want to see what the current subject is, you can issue the *@TS1?* command.

Sending E-Mail

There are a few ways to send e-mail using the iModem. The simplest way is to assert a special pin on the iModem. This causes a predefined e-mail message to be sent. Message 0 is a predefined message that reports the status of two uncommitted inputs on the iModem. By using this mode, you can use the iModem to report the status of several inputs, and you might not need a microprocessor at all. You can also send a special command (*@TDM0*) to force the sending of the same e-mail. You can also set the modem to send a custom message when you assert the send pin by sending *@TJ1*. Usually, if you are using a microprocessor, you'll want to send the customized e-mail message. You can do this via the *@TM1* command, which sets the message text, and the *@TDM1* command, which actually sends the message.

While sending mail, there is a stream of progress messages to indicate the current status of the operation. The typical sequence is as follows:

```
CONNECT
PASSWORD OK
MESSAGE ACCEPTED
HANGING UP
NO CONNECT
```

Sending commands include the following:

@TD	Send default message (same as *@TDM0*).
@TDM0.	Same as *@TD*.
@TDM1.	Send custom message.
@TM1.	Set custom message.
@TS1.	Set subject for outgoing e-mail (15 character mixed).

Using the Javelin

Listing 9-1 shows a simple Javelin program that sends a message. The text of the message shows the current temperature (from a DS1620 temperature sensor).

Listing 9-1
Sending E-Mail

```
import stamp.core.*;
import stamp.peripheral.sensor.temperature.DS1620;


/**
 * Handler for iModem
 * @version 1.0
 * @author Al Williams
 */
```

```java
public class iModemSend {

  final static int SERIAL_RX_PIN = CPU.pin0;
  final static int SERIAL_TX_PIN = CPU.pin1;
  static Uart rxUart = new Uart( Uart.dirReceive, SERIAL_RX_PIN,
                Uart.dontInvert,
                Uart.speed2400,
                Uart.stop1 );
  static Uart txUart = new Uart( Uart.dirTransmit, SERIAL_TX_PIN,
                Uart.dontInvert,
                Uart.speed2400,
                Uart.stop1 );

static StringBuffer t=new StringBuffer(81);
static Timer waittimer = new Timer();

public static void sendString(String s) {
  for (int i=0;i<s.length();i++) {
    txUart.sendByte(s.charAt(i));
    CPU.delay(500);      // modem needs delay between characters
    }
}

public static boolean waitOK() {
  return waitOK(2);
  }

public static boolean waitOK(int timeout) {
  Timer to=waittimer;
  char c=' ';
  int cnt=0;
  to.mark();
  do {
    if (rxUart.byteAvailable()) {
      c=(char)rxUart.receiveByte();
      System.out.print(c);
      if (cnt==0 && c=='O') cnt++;
      if (cnt==1 && c=='K') cnt++;
      if (cnt==2 && c=='\r') cnt++;
```

Listing 9-1

Continued

```
         }
      } while (cnt!=3 && !to.timeoutSec(timeout));
      if (cnt!=3) System.out.println("Timeout!");
   return cnt==3;
   }


   public static boolean waitCON() {
    Timer to=waittimer;
    char c=' ';
    int cnt1=0,cnt2=0;
    to.mark();
    do {
       if (rxUart.byteAvailable()) {
         c=(char)rxUart.receiveByte();
         System.out.print(c);
         if (c==0) continue;
// look for Accepted
         if (cnt1==0 && c=='E') { cnt1++; continue; }
         if (cnt1==1 && c=='C') { cnt1++; continue; }
         if (cnt1==2 && c=='T') {
            while ((c=(char)rxUart.receiveByte())!='\r')
              System.out.print(c);  // wait for EOL
            System.out.print('\r');
            return true;
            }
         cnt1=0;
// look for No connect
         if (cnt2==0 && c=='N') { cnt2++; continue; }
         if (cnt2==1 && c=='O') { cnt2++; continue; }
         if (cnt2==2 && c==' ') {
            while ((c=(char)rxUart.receiveByte())!='\r')
              System.out.print(c);  // wait for EOL
            System.out.print('\r');
            return false;
            }
         cnt2=0;
         }
      } while (!to.timeoutSec(90));
      System.out.println("Timeout");
```

Listing 9-1
Continued

```
      return false;
  }

  public static boolean waitTED() {
   Timer to=waittimer;
   char c=' ';
   int cnt1=0,cnt2=0;
   to.mark();
   do {
     if (rxUart.byteAvailable()) {
       c=(char)rxUart.receiveByte();
       System.out.print(c);
       if (c==0) continue;
// look for Accepted
       if (cnt1==0 && c=='T') { cnt1++; continue; }
       if (cnt1==1 && c=='E') { cnt1++; continue; }
       if (cnt1==2 && c=='D') { return true; }
       cnt1=0;
// look for No connect
       if (cnt2==0 && c=='N') { cnt2++; continue; }
       if (cnt2==1 && c=='O') { cnt2++; continue; }
       if (cnt2==2 && c==' ') {
          while (rxUart.receiveByte()!='\r');  // wait for EOL
          return false;
          }
       cnt2=0;

       }
     } while (!to.timeoutSec(90));
     System.out.println("Timeout");
     return false;
  }

public static boolean findModem() {
  sendString("\rAT\r");
  if (!waitOK()) {
    System.out.println("Can't find modem");
    return false;
```

Listing 9-1
Continued

```
      }
    return true;
    }
    public static void main() {
    // careful... the iModem takes a few seconds to "wake up"
     Timer clock=new Timer();
     Timer level0timer = new Timer();
     DS1620 sensor=null;
     int cnt=-1;
     int temp;
     int i;

     sensor=new DS1620(CPU.pin4,CPU.pin5,CPU.pin6);

     // let's try to see if the modem is awake and clear
     clock.mark();
    out:
      do {
        txUart.sendByte('\');
        do {
          if (rxUart.byteAvailable() && rxUart.receiveByte()=='\r')
            break out;
          } while (!clock.timeout(4000));
        } while (true);

    mainloop:
      while (true) {
        if (rxUart.byteAvailable())
          System.out.print((char)rxUart.receiveByte());
        // only check every 30s
        if (cnt!=-1 && !clock.timeoutSec(30)) continue;
        clock.mark();
        cnt++;
        if (cnt%10!=0) continue;  // only every 5 minutes
        cnt=0; // stop future roll over
        temp=sensor.getTempF();
        t.clear();
        t.append(Integer.toString(temp));
        if (!findModem()) continue; // no modem
```

Listing 9-1
Continued

```
        do
            sendString("@TS1=Temp Report\r");
        while (!waitOK());
        do
            sendString("@TA1=alw@al-williams.com\r");
        while (!waitOK());
        do {
            sendString("@TM1=Greetings from Javelin!\r");
            CPU.delay(100);   // pause for message line processing
            sendString("The temperature is ");
            CPU.delay(100);
            sendString(t.toString());
            CPU.delay(100);
            sendString("\r");
            CPU.delay(100);
            sendString(".\r");
        } while (!waitOK());
        CPU.delay(250);
        sendString("@TDM1\r");
        if (!waitTED()) {
            System.out.println("Can't connect!");
            }
        System.out.println("Done");
        }
    }
}
```

There are a few points to pay special attention to in Listing 9-1. First, notice that the *sendString* method inserts a delay between each character, as the modem requires. The *findModem* method uses the normal *AT* command to elicit an *OK* response from the modem. This ensures that the modem is powered on and in the command state.

When power is applied to the entire system, the modem takes a bit longer to wake up than a typical processor. For that reason, the first part of the program sends a carriage return until it sees an echo from the modem. This is in addition to the *findModem* test.

The program makes extensive use of the Javelin's *Timer* object. This virtual peripheral is unique since only the first instance actually consumes a virtual peripheral slot. Subsequent instances piggyback on the original instance. The program uses timers to control the frequency of e-mail transmissions. It also uses a timer to prevent the program from waiting forever for a modem response that doesn't occur.

The order of commands is:

@TS1.	Set subject.
@TA1.	Set address.
@TM1.	Set message.
@TDM1.	Send message.

Setting the message with *@TM1* requires you to send a carriage return at the end of each line. The last line must be a period followed by a carriage return. A *CONNECTED* message indicates success.

Receiving E-Mail

Using the same framework, it is relatively straightforward to receive e-mail addressed to the modem. You can see an example in Listing 9-2. The *@TDG1* command initiates an e-mail retrieval. The normal responses are as follows:

Listing 9-2
Receiving with
iModem

```
import stamp.core.*;
import stamp.peripheral.sensor.temperature.DS1620;

/**
 * Handler for iModem
 * @version 1.0
 * @author Al Williams
 */

public class iModemTest {

    final static int SERIAL_RX_PIN = CPU.pin0;
```

Listing 9-2

Continued

```
   final static int SERIAL_TX_PIN = CPU.pin1;
// final static int SERIAL_CTS_PIN = CPU.pin2;
// final static int SERIAL_RTS_PIN = CPU.pin3;

  static Uart rxUart = new Uart( Uart.dirReceive, SERIAL_RX_PIN,
                 Uart.dontInvert,
                 Uart.speed2400,
                 Uart.stop1 );
  static Uart txUart = new Uart( Uart.dirTransmit, SERIAL_TX_PIN,
                 Uart.dontInvert,
                 Uart.speed2400,
                 Uart.stop1 );
static StringBuffer line = new StringBuffer(256);
static StringBuffer t=new StringBuffer(81);
static int msgct;
static Timer waittimer = new Timer();

public static void sendString(String s) {
   for (int i=0;i<s.length();i++) {
     txUart.sendByte(s.charAt(i));
     CPU.delay(500);     // modem needs delay between characters
     }
}

public static boolean getLine() {
   return getLine(90);
   }

public static boolean getLine(int timeout) {
   Timer to=waittimer;
   char c=' ';
   line.clear();
   to.mark();
   do {
     if (rxUart.byteAvailable()) {
       to.mark();  // reset timer
       c=(char)rxUart.receiveByte();
       System.out.print(c);
       if (c!='\r' && c!='\n') line.append(c);
```

```
      }
    } while (c!='\n' && !to.timeoutSec(timeout));
    if (c!='\n') return false;   // timed out!
    if (line.toString().equals("NO CONNECT")) {
      System.out.println("Early drop out!");
      return false;
      }
    return true;
    }

public static boolean waitOK() {
  return waitOK(2);
  }

public static boolean waitOK(int timeout) {
Timer to=waittimer;
 char c=' ';
 int cnt=0;
 to.mark();
 do {
   if (rxUart.byteAvailable()) {
     c=(char)rxUart.receiveByte();
     System.out.print(c);
     if (cnt==0 && c=='O') cnt++;
     if (cnt==1 && c=='K') cnt++;
     if (cnt==2 && c=='\r') cnt++;
     }
   } while (cnt!=3 && !to.timeoutSec(timeout));
   if (cnt!=3) System.out.println("Timeout!");
return cnt==3;
}

public static boolean waitCON() {
 Timer to=waittimer;
 char c=' ';
 int cnt1=0,cnt2=0;
 to.mark();
 do {
```

Listing 9-2

Continued

```
         if (rxUart.byteAvailable()) {
           c=(char)rxUart.receiveByte();
           System.out.print(c);
           if (c==0) continue;
// look for Accepted
           if (cnt1==0 && c=='E') { cnt1++; continue; }
           if (cnt1==1 && c=='C') { cnt1++; continue; }
           if (cnt1==2 && c=='T') {
             while ((c=(char)rxUart.receiveByte())!='\r')
               System.out.print(c);  // wait for EOL
             System.out.print('\r');
             return true;
             }
           cnt1=0;
// look for No connect
           if (cnt2==0 && c=='N') { cnt2++; continue; }
           if (cnt2==1 && c=='O') { cnt2++; continue; }
           if (cnt2==2 && c==' ') {
             while ((c=(char)rxUart.receiveByte())!='\r')
               System.out.print(c);  // wait for EOL
             System.out.print('\r');
             return false;
             }
           cnt2=0;

           }
       } while (!to.timeoutSec(90));
       System.out.println("Timeout");
       return false;
   }

  public static boolean waitTED() {
   Timer to=waittimer;
   char c=' ';
   int cnt1=0,cnt2=0;
   to.mark();
   do {
     if (rxUart.byteAvailable()) {
       c=(char)rxUart.receiveByte();
```

Listing 9-2
Continued

```
        System.out.print(c);
        if (c==0) continue;
// look for Accepted
        if (cnt1==0 && c=='T') { cnt1++; continue; }
        if (cnt1==1 && c=='E') { cnt1++; continue; }
        if (cnt1==2 && c=='D') { return true; }
        cnt1=0;
// look for No connect
        if (cnt2==0 && c=='N') { cnt2++; continue; }
        if (cnt2==1 && c=='O') { cnt2++; continue; }
        if (cnt2==2 && c==' ') {
          while (rxUart.receiveByte()!='\r');  // wait for EOL
          return false;
          }
        cnt2=0;

        }
    } while (!to.timeoutSec(90));
    System.out.println("Timeout");
    return false;
  }

  public static char linecharAt(int n) {
    try {
      return line.charAt(n);
      }
    catch (IndexOutOfBoundsException e) {
      return '<\\>0';
      }
    }

  public static boolean readNumMsg() {
    int i;
    if (!getLine()) return false;
    if (line.equals("NO MAIL")) {
      msgct=0;
      return true;
      }
```

Listing 9-2

Continued

```
      if (linecharAt(0)<'0' || linecharAt(0)>'9') return false;  //
huh?
      // this tells us how many messages are in
      t.clear();
      i=0;
      do {
        if (i>line.length()) break;
        if (linecharAt(i)==' ' || linecharAt(i)=='\0') break;
        t.append(linecharAt(i));
        i++;
        } while (true);
        msgct=Integer.parseInt(t);
      return true;
      }


public static boolean findModem() {
  sendString("\rAT\r");
  if (!waitOK()) {
    System.out.println("Can't find modem");
    return false;
    }
  return true;
  }


public static boolean delmsg() {
      if (!findModem()) return false;
      // delete it
      sendString("@TDK1\r");
      if (!waitCON()) {   // wait for CONNECT
        System.out.println("Delete failed");
        return false;
        }
      if (!waitCON()) {   // wait for POP3 CONNECT
        System.out.println("Delete failed");
        return false;
        }
      if (!readNumMsg()) {
          System.out.println("Did not find message count");
          return false;
```

Listing 9-2
Continued

```
            }
        waitCON();  // wait for NO CONNECT
            return true;
            }

    public static void main() {
    // careful... the iModem takes a few seconds to "wake up"
    Timer clock=new Timer();
    Timer level0timer = new Timer();
    DS1620 sensor=null;
    int cnt=-1;
    int temp;
    int i;

    sensor=new DS1620(CPU.pin4,CPU.pin5,CPU.pin6);

    // let's try to see if the modem is awake and clear
    clock.mark();
    out:
      do {
        txUart.sendByte('\r');
        do {
          if (rxUart.byteAvailable() && rxUart.receiveByte()=='\r')
break out;
          } while (!clock.timeout(4000));
        } while (true);

    mainloop:
      while (true) {
        if (rxUart.byteAvailable())
System.out.print((char)rxUart.receiveByte());
        if (cnt!=-1 && !clock.timeoutSec(30)) continue;  // only
check every 30s
        clock.mark();
        cnt++;
        if (cnt%10!=0) continue;  // only every 5 minutes
        cnt=0; // stop future roll over
        temp=sensor.getTempF();
        t.clear();
        t.append(Integer.toString(temp));
```

Listing 9-2
Continued

```
if (!findModem()) continue; // no modem
do
  sendString("@TS1=Temp Report\r");
while (!waitOK());
do
  sendString("@TA1=wd5gnr@yahoo.com\r");
while (!waitOK());
do {
  sendString("@TM1=Greetings from Javelin!\r");
  CPU.delay(100);  // pause for message line processing
  sendString("The temperature is ");
  CPU.delay(100);
  sendString(t.toString());
  CPU.delay(100);
  sendString("\r");
  CPU.delay(100);
  sendString(".\r");
} while (!waitOK());
CPU.delay(250);
sendString("@TDM1\r");
if (!waitTED()) {
    System.out.println("Can't connect!");
    }
do {
  sendString("@TDG1\r");
  if (!waitOK(60)) {   // wait for password OK
    System.out.println("Can't connect!");
    continue mainloop;
    }
  if (!waitCON()) {  // wait for POP3 Connect
    System.out.println("Can't connect to POP3");
    continue mainloop;
    }
  if (!readNumMsg()) {
    System.out.println("Didn't find number of messages");
    continue mainloop;
    }
```

Listing 9-2
Continued

```
            if (msgct==0) {
                System.out.println("No mail!");
                }
            else
                {
                if (!getLine()) continue mainloop;
        // Now date, From, Subject - - warning: sometimes the subject
        // is mangled by the ISP so best not to rely on it
                if (!getLine()) continue mainloop;
    // get message body
                do {
                if (!getLine()) continue mainloop;
                } while (linecharAt(0)!='.' || linecharAt(1)!='\0');
            // process this e-mail
            System.out.println("***Email received");
            // in real life you'd have done something
            //with the input lines
            }
            // wait for NO CONNECT (will return false but that's OK)
            waitCON();
            if (msgct!=0) {
            level0timer.mark();
            // wait a good bit seconds while (!level0timer.timeout-
Sec(300));
            while (!delmsg()) {
            level0timer.mark();
            // keep trying to delete until it works
            while (!level0timer.timeoutSec(150));
            }
        }
    } while (msgct!=0);
  System.out.println("Done");
 }
 }


 }
```

```
CONNECTED
PASSWORD OK
POP3 CONNECT
2 MESSAGES
```

Following this preamble is the text message. Each time you retrieve a message, the modem dials the ISP. Also, retrieving a message does not delete it. You can delete messages with *@TDK1*. When you delete message 1, all the subsequent messages renumber themselves, so the first pending message is always message 1.

If there are no messages pending, you don't get a *0 MESSAGES* string as you might expect. Instead, the modem reports *NO MESSAGES*, which makes it a bit harder to parse the responses. The end of the incoming message uses the same period on a single line that you use to indicate the end of a sent message.

Limitations

The custom outbound message is limited to 100 characters (including blanks and spaces). The subject line has a limit of 15 characters. There is a similar limit to incoming messages. However, the biggest problem with the receiving program is that the modem (or perhaps the ISP) doesn't properly disconnect. So when you retrieve or delete a message, subsequent attempts to access the POP3 server will fail until the server resets. There is no good way around this—you simply have to check for failure and retry until the operation successfully completes.

Using the ISP

The Press4Service ISP provided by Cermetek (for a monthly fee) allows you to transform e-mail into different types of data. The ISP service decodes incoming mail addresses according to the following format:

```
mailbox.service@alanet.net
```

The mailbox identifies a particular user ID assigned by the ISP. However, the service portion can be any of the following: e-mail, fax, voice, Web, or FTP. Notice that the fax and voice services are not included in the current Cermetek free trial of Press4Service.

You can log into the Press4Service database using an ordinary Web browser and set up different options. For example, the e-mail service can forward mail to any e-mail addresses you like. For the fax and voice services, you can specify a destination phone number.

The Web service stores the incoming e-mail as an ASCII file. The file's name will correspond to the modem's serial number with an extension of .txt. It is possible to read this file into a dynamic Web page hosted by Press4Service and either display it totally or parse information from it and display it on the Web. If you prefer, you can use the FTP service to send the e-mail to a remote server of your choice for further processing.

Of course, you can easily write a Java program, for instance, that does custom processing of incoming e-mail if you have your own mail infrastructure. However, if you don't have flexible e-mail, the Press4Service ISP can fill the gap.

CHAPTER 1

The Once and Future Internet

It is hard to imagine that the Internet grew from a largely unknown network connecting universities and government sites. Now it seems that the Internet is everywhere, but it hasn't been that way very long. I suspect that new wireless options and technological advances will at some point make the Internet (or something like it) even more ubiquitous than it already is.

By that time, adding Internet connectivity to a product will be much simpler than it is today. It wasn't that long ago that adding, say, an LCD display to a microcontroller required a great deal of specialized knowledge and coding. Now, predesigned modules make it simple for even the smallest microcontrollers to drive an LCD. The same thing will eventually become true of the Internet. At some time in the future, practically all processors will have the power (and the vendor support) to handle an Internet connection.

Today, that isn't the case. Sure, larger processors can handle a network connection (assuming you can make the physical connection). However, the smallest processors, often used in high-volume production, don't have enough resources to handle networking and the tasks they are meant to handle at the same time. Also, many legacy devices don't have any support for a network connection.

Until this rosy future, there will be a market for bridge devices that connect between classic embedded systems and the network. You can roll your own using a device like the TINI, of course. There are also no shortage of devices (like the iModem) that can provide a ready-made interface.

The Right Tools for the Job

It is very tempting to slap a Web browser into every embedded system. This temptation grows even stronger when the processor can easily handle the software required. In some cases, a Web interface is a good idea. For example, most network routers (which are nothing more than embedded systems) contain a Web server. The end user then uses a normal Web browser to configure the router.

However, for most embedded systems, you'll probably want to avoid a straight Web browser interface. Consider that an embedded system obviously has some task to do. Every cycle you spend listening for Web requests, encoding data, and processing pages is another cycle you aren't handling your main task. Working with network packets is asynchronous in nature, so real-time processing may be hindered. Packet fragmentation requires the TCP/IP stack to buffer partial packets, which can be very taxing on small systems with limited memory.

For many small systems, a UDP solution works well, but you can't connect to a Web browser using UDP. A custom client can be useful, or you may want to use a gateway program on an Internet server that translates between the UDP packets and more conventional protocols.

If you really have to have a Web interface, you might consider a separate controller just to handle the network interface. You could roll your own interface, using something like a TINI, or you could use one of many commercial products that offer this type of function. As an alternative, you could roll your own with software products (generally in C) that add a TCP/IP stack to the microcontroller of your choice.

Of course, another option is to use a PC or one of the many PC/104 boards that are available. With that much horsepower, it is easy to run a full-blown networking stack. Most operating systems targeted at embedded PCs will provide networking support out of the box.

If you do want to port a stack or use a commercial device, you might consider some of these offerings:

Allegro RomPager: http://www.allegrosoft.com. Allegro's RomPager is an embeddable stack and Web server that you can target for many different embedded systems. RomPager has its own scheduler and allows you to define parameters to set or get via the Web, Simple Network Management Protocol (SNMP), or telnet.

Arcturus uCdimm: http://www.arcturusnetworks.com. The uCdimm offering from Arcturus is a tiny embedded computer in an soDIMM form factor. It has TCP/IP, Ethernet, and serial support.

Bagotronix DOS Stamp: http://www.bagotronix.com. Bagotronix's DOS Stamp has an optional TCP/IP stack (DIME) that you can install to turn it into a Web server. It can also handle e-mail. The DOS Stamp uses a serial line and can handle SLIP or PPP.

Beck IPC IPC@Chip: http://www.bcl-online.de. This German company makes an embedded Web server in a chip-like module. The device can directly attach to Ethernet.

Cermetek iModem: http://www.cermetek.com. In Chap. 9, I showed you the Cermetek iModem. Depending on your point of view, the iModem is either an e-mail sending and receiving engine, or a general gateway (if you use the Cermetek ISP). Although it seems more limited than a general-purpose Web server, as I pointed out in Chap. 9, e-mail has certain benefits for many embedded systems.

EBSNet RTip: http://www.ebsnetinc.com/.   The RTip stack is a C language stack that is meant for embedded processors. It supports most major protocols and has an optional Web server.

JK Microsystems: http://www.jkmicro.com.   JK makes a variety of enclosed and board-level products that have TCP/IP stacks.

Lantronix: http://www.lantronix.com/.   The Lantronix offerings include board-level and chip-level products. The boards accept TTL-level RS232s and convert to Ethernet via an internal Web server. The chip has several possible interfaces (RS232, CAN, SPI, I2C, and USB).

LiveDevices Ebedinet: http://www.livedevices.com.   LiveDevices offers a TCP I/P stack for several processors including the Microchip PIC (18x series). It even has a live camera demo where you can change the state of some LEDs on a board and see the result.

NetBurner: http://www.netburner.com.   NetBurner is a board-level product that offers an Ethernet connection along with RS232, SPI, and RS485/RS422 interfaces. The board has a simplified mode of operation, or you can develop custom programs using the GNU C++ compiler included with the development kit.

NetMedia SitePlayer: http://www.siteplayer.com.   NetMedia's tiny SitePlayer module is a Web server and Ethernet adapter that can communicate with a microcontroller via a serial port or a synchronous Serial Peripheral Interface (SPI) -like interface.

Rabbit Semiconductor: http://www.rabbitsemiconductor.com/.   The popular Rabbit computer boards have a royalty-free TCP/IP stack. The boards are also available with Ethernet connectors, so you can use them to create a customized gateway.

Sena HelloDevice: http://www.sena.com.   Sena makes the oddly named HelloDevice, which can convert a variety of serial protocols into Ethernet. One way to use this product is with Serial/IP (from Tactical Software). Combined with a HelloDevice, a PC can drive a serial device over the Internet. The HelloDevice is available in a box, as a board-level product, or as an IC-like module.

Sumbox: http://www.sumbox.com. Sumbox makes several small computers that are Ethernet-capable. In addition, the company makes an Ethernet modem that allows an embedded computer to talk to the network as if it were a standard modem. For example, the ATD command connects to a particular IP address (and port). ATS0=1 will allow the modem to accept incoming requests (answer mode).

Looking Forward

The Internet is constantly growing and changing. As I write this, there are two technologies on the horizon that aren't very important today, but could have an impact on embedded systems in the near future: IPV6 and multicasting.

It doesn't take a math genius to realize that a standard IP address (like 140.99.17.25) has only 32 bits in it and therefore there are only around 4 billion IP addresses possible. In reality, the actual number of addresses is less than you'd expect since certain addresses are reserved and not generally usable. IPV6 addresses this problem by allowing, among other things, a larger space for IP addresses.

Multicasting is a way to make broadcasting data on the Internet more efficient. Suppose you have a sensor that is monitoring a plastic plant outside of Chicago. Ten people in Houston and three people in Los Angeles want to monitor the data. The poor embedded system has to send packets to 13 different destinations. This is true even though most of the packets will traverse identical network routes on their way to Houston and the rest will probably go the same route to Los Angeles.

With multicasting, special routers can determine that the same data stream is destined for multiple recipients and arrange for a single broadcast to that router. The router then handles duplicating the data for each recipient (some of these recipients may, in fact, be other multicasting routers).

About IPV6

The most obvious feature of IPV6 (when compared to the current version of IP, which is version 4) is the increased space for IP addresses. While normal IP uses 32-bit addresses, IPV6 allows up to 128 bits for an address.

Although 4 billion addresses sounds like a lot, there are several reasons why the space is inadequate. At first, the Internet designers divided the addresses into three classes of networks. A class A network, for example, allowed one network to contain 16 million addresses! Needless to say, there aren't many networks that need that many addresses, so many class A addresses are unused (but also mostly unusable). Also, as mentioned before, many of the technically possible addresses aren't legal anyway (for example, loop-back addresses and broadcast addresses).

With the explosive growth of the Internet, it looked like a certainty that the Internet would exhaust the available IP addresses—and possibly before it would be feasible to adopt IPV6. Luckily, CIDR (classless interdomain routing, sometimes known as supernetting) made it possible to reuse many of the old class-style addresses that were not being used.

However, even this is a stopgap measure. As more devices (including embedded devices) require IP addresses, it is just a matter of time before there are no more left. However the 128-bit address space provides a mind-boggling number of addresses (about 3.4×10^{38} according to my calculator).

IPV6 also makes two other improvements to regular IP: security and roaming. With today's IP scheme, it is difficult for a device to seamlessly change IP addresses. This is important if you are, for example, using a cell phone and driving between different network areas (or perhaps if a robot is using some wireless network technology).

Security means that all IPV6 traffic can be encrypted. You can do this with IPSec now in some TCP/IP stacks, but an IPV6 stack must support encryption.

Although IPV6 addresses are large, they are still segregated into groups much like the old-style IP addresses (before CIDR). Typically, the first 48 bits of the address indicate a root-level provider. The next 16 bits are a subnetwork number. The remaining 64 bits indicate the specific host on the network. Curiously enough, Ethernet hardware addresses are only 48 bits wide, so by using an address derived from the Ethernet hardware address, the network can configure itself without resorting to Dynamic Host Control Protocol (DHCP).

Multicasting

Although many people look to multicasting as a way to replace things like conventional television and radio broadcasting, I think it has tremendous potential for embedded systems that do things like data collection. The idea is simple: Instead of sending data to a specific IP address, multicasting clients send their data to a special multicast IP address. This address is really handled by a multicast router.

Routers that handle multicasting support IGMP (Internet Group Management Protocol). The problem is, there isn't widespread support for this on the general Internet (although on an intranet you may be surprised to find you have multicasting capability and don't even know it).

Multicasting can be especially important when underpowered embedded systems with poor network links need to send data to many consumers. As long as the multicasting router is well connected, you could handle a virtually unlimited number of clients using multicasting techniques.

What's Next

In 1965, Gordon Moore observed that the number of transistors on an IC doubled every few years. The IBM PC's original 8088 processor had around 30,000 transistors on the die while today's Pentium IV has over 40,000,000! Granted, that's a desktop processor, but it is clear that today's microcontrollers are significantly more powerful than their predecessors.

In the foreseeable future, simple microcontrollers will easily handle networking tasks that are taxing to today's processors. Not only will increased memory and computing power be helpful, but dedicated peripherals and architectures built for networking will make this a reality.

The network connection itself is another area that will change dramatically. Over the years, networking has moved from traditional Ethernet, to phone lines, to thin and telephone wire-style Ethernet. Today, wireless networks like 802.11 are bringing the Internet to more corners of the world than ever before.

As it becomes more important for embedded systems to connect to the network, you'll see much more hardware support for connections to the network. We may even see special lightweight networking made available for small mobile devices.

There was a time when building an embedded system to remotely monitor a temperature was very complicated. Today you can buy an off-the-shelf modem, a Basic Stamp, and a smart temperature sensor with a serial interface and have the whole system complete in a few hours. In 10 years you'll be able to knock together a networked embedded system just as easily. For today, though, you'll have to stick to the techniques covered in this book. Good luck!

Appendix

INSIDE PPP

Point to Point Protocol (PPP) is defined in STD051 (one of the Internet standard documents). It is also known as RFC 1661. PPP defines three main components:

1. A method for encapsulating datagrams.

2. A link control protocol (LCP) that allows control of the PPP connection.

3. A family of network control protocols (NCPs) for configuring network-layer protocols.

The PPP encapsulation is quite simple. Each packet has a one- or two-byte protocol field. Next, the packet contains the data you wish to send. Finally, there may be some padding bytes to make the packet a certain size if necessary.

All protocol bytes are odd numbers (to comply with ISO3309). Also, if the protocol number requires 2 bytes, the most significant byte must be even. This allows you to easily determine if the protocol number is 1 byte or 2 bytes without having to maintain a list of specific protocol numbers.

The first 4 bits of the protocol number determine the type of protocol. Specifically,

0xxx—3xxx	Network layer protocol
4xxx—7xxx	Low-volume protocols with no NCP
8xxx—Bxxx	NCP packets
Cxxx—Fxxx	LCP packets

The standard reserves several protocol numbers:

0001	Padding
007D	Control escape
8021	IPCP (IP Control Protocol)
C021	LCP (Link Control Protocol)
C023	Password Authentication Protocol (PAP)
C025	Link quality report
C223	Challenge Handshake Authentication Protocol (CHAP)

There are also several reserved numbers that are not commonly used. Although not defined in the standard, the protocol number of most interest to you is 0021, which is the number for IP. PPP can encapsulate many different kinds of network traffic, and you can find a complete list in STD2. For Internet connections, IP is the protocol you'll want to carry via PPP.

The information field contains the actual bytes you want to send via PPP. The Maximum Receive Unit (MRU) determines how long this field can be. By default, the MRU is 1500 bytes, but the protocol allows both parties to negotiate a different MRU if they desire.

In addition, the transmission may add extra padding bytes to a packet if it wants to round the packet size to a certain size (up to the MRU). For example, the sender may wish always to send an even number of bytes and will add an extra byte to odd-sized packets.

State Diagram

The typical PPP state diagram appears in Fig. A-1. The initial PPP state is "dead." When you want to establish the link, you follow the "up" arrow to the "establish" state.

During the "establish" state, the PPP software handles LCP packets to negotiate PPP options. Options for specific network protocols are han-

Figure A-1
Point-to-Point
Protocol States

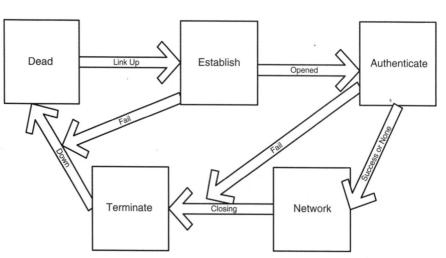

dled separately. Any non-LCP packets received during this phase must be discarded without warning or error messages. The state may return to "establish" when the program receives an LCP configure request in later states (such as authentication).

The "authentication" phase is not required. In systems that require an external log-in, you may not require the user to provide a user ID and password. However, direct PPP connections will almost certainly require a user ID and password.

There are several different ways to transmit the user ID and password. The basic method—PAP, or Password Authentication Protocol—sends the information in the clear. However, there are other methods that require sending encrypted data or using the password to encode a challenge string. These are more secure but also more difficult to implement.

Once authentication is complete (or determined to be unnecessary) the protocol is free to send and receive network protocol packets. NCP packets can provide specific configuration (for example, the PPP host might inform the client what IP address it should use).

The final state is "terminate," where one side sends the other side an LCP packet requesting a termination. This implies the termination of the entire link. Technically, PPP can close a network protocol without dropping the PPP link, although for Internet purposes, this is probably not very useful.

Options

During the link establishment phase, all options are set to default values. However, either side of the conversation can request a change in options. This is done via LCP packets. Any non-LCP packets received are silently discarded.

The side that wishes to set options will send a Configure Request packet (LCP code 1). The first byte of the packet is the code and the following byte is a unique identifier the sender uses to distinguish this request from others. Next, the packet contains a two-byte length. This length is the size, in bytes, of the entire packet (including the code field, the identifier field, and even the length field).

Following this fixed-format header, the packet contains any number of configuration options requested. Each configuration option consists of a code number, a length in bytes, and a varying number of data bytes. The length is the length of the entire configuration sequence including the code number and length bytes.

There are six option codes defined:

1. Maximum receive unit (MRU)

2. Authentication protocol

3. Quality protocol

4. Magic number

5. Protocol field compression

6. Address and control field compression

This MRU allows either side to limit the maximum size of sent packets. By default, the MRU is 1500 bytes. Of course, packets can be smaller than the MRU. Although it might be tempting to set the MRU to a smaller number for an embedded system, you may have to receive 1500-byte packets before options are negotiated, so it is customary to allow for reception of 1500 bytes. However, a small processor like the Javelin may have to take the chance (the Javelin's PPP code sets an MRU of 254 bytes, for example). The length of an MRU option is 4 bytes, so to request an MRU of 2047 bytes (7FF hex), for example, the packet would look like this:

```
01 80 00 08 01 04 07 FF
```

Here the unique ID is hex 80. The length of the entire LCP packet is 8 bytes. The configuration packet's length is 4 bytes.

The authentication protocol indicates what protocol is required for authentication. Typically, dialup users authenticate with the server, but not vice versa. However, it is possible to have both sides authenticate if that is desirable. The length of an authentication protocol packet varies since custom protocols might include data that is specific to the protocol.

The default, of course, is to do no authentication. To request authentication, the data field will contain a 2-byte identifier specifying the protocol. This is often C023 (PAP) or C223 (CHAP). The PAP protocol simply sends the user ID and password in the clear. This has the unfortunate effect that anyone listening to the line can extract the log-in information. CHAP sends a random challenge. The user's computer then applies a special algorithm to the challenge that uses the password. It then sends the result to the host computer. The host computer computes the same algorithm and the results must match for authentication to succeed. The algorithm is such that it would be difficult to guess the password by examining the challenge and the response. The challenge changes each time, so simply knowing the response isn't sufficient to break into the host computer.

The other configuration options are often ignored by small embedded computers. The quality protocol option allows the computers to send each other information regarding how robust the link between them is. The magic number request allows a PPP program to determine if it is, in fact, talking to itself. The compression options allow you to omit certain repetitive information in packet headers. However, this is not compression of the data. It simply allows you to omit certain header fields that typically have the same value on every packet, anyway.

When a PPP implementation receives a configuration request, it can reply in one of three ways. First, if any options are unknown or non-negotiable, the program replies with a configure reject (LCP code 4). This packet has the code number, the identifier from the original request, a 2-byte length field, and then a list of all the options that are rejected. This list is in the same format that the configure request uses. In fact, the options must be in the same order they appear in for the configure request. However, the only options that appear are the ones that are rejected.

Notice that rejecting means you can't even consider the option. If you simply don't like the choices offered, that isn't a rejection. So, for example, if the host wants to set the MRU to 4096 and you want a different value, you don't reject the request (instead, you use a negative acknowledge as you'll see shortly).

When the requestor receives the rejection, it will try again after removing the unacceptable options. If any of the options are not acceptable (for example, the MRU is too high), the receiver replies with a configure negative acknowledge (LCP code 3). This response is similar to the reject response, except that the list now contains the options with unacceptable values modified to reflect values that would be acceptable.

Suppose that a PPP stack receives the following configure request:

```
01 80 00 08 01 04 07 FF
```

This is a request for a 2047-byte MRU. If the stack wants an MRU of 240 bytes (F0 hex) it would reply:

```
03 80 00 08 01 04 00 F0
```

Again, the only options in the list are the ones in dispute and they must appear in the same order that the original configure request sent them. The sender will then try again with different values.

The third case is the one in which the configure request contains only options and values that are acceptable. In this case, the response is a configure acknowledge (LCP code 2) that contains the exact same option list as the configure request. So if the PPP stack wanted to accept the previous 2047-byte MRU, it would reply

```
02 80 00 08 01 04 07 FF
```

The other LCP packets that you'll commonly encounter are the ones to request a termination (5) or acknowledge termination (6); reject an unknown LCP packet type (7); reject an unknown protocol type (8); or request or provide an echo (9 and 10).

HDLC Framing

When transmitted over a modem, the PPP packets are usually encapsulated using HDLC-like framing (HDLC stands for High-Level Data Link Control, and is a common method of transmitting data synchronously). This is defined in RFC1662.

The HDLC framing makes it possible to identify packet boundaries in a stream of data. To do this, each HDLC frame starts with 7E FF 03 followed by the data of the PPP packet. Following the packet is a 16-bit Frame Check Sequence (FCS) that is computed on the packet. Obviously, the receiver should be able to recompute the FCS and arrive at the same answer as the sender. There is an optional 32-bit FCS that can be negotiated, if desired. The packet may end with a 7E byte which may start the next frame, if desired.

To ensure that the 7E doesn't appear in the data stream, the protocol requires an escape character (7D). The escape character indicates that the next character requires an exclusive or operation with 20 hex to decode it. At a minimum, you must escape 7E and 7D characters. So to send a 7D, you must actually send 7D 5D. It is customary to escape control characters (those less than 20 hex). So to send a 3, for example, you'd send 7D 23. This applies even to the 3 in the HDLC header. Using this escape mechanism prevents communications channels from interfering with data by inserting or removing control characters.

The FCS is computed on the decoded data, not the escaped data. So for the purposes of FCS computation, the 7D 5D sequence is simply 7D. You can find the exact algorithm employed in RFC1662.

Beyond PPP

You can think of PPP as sort of a ministack that resides within the actual stack's protocol layer. At the bottom of this pseudostack is the HDLC framing. Above this layer is the PPP encoding. The protocols that you send—IP, LCP, and IPCP—use these two sublayers to communicate. Just as LCP configures the PPP connection, IPCP configures the IP protocol.

IPCP works very much like LCP. Typically, the embedded system will send an option request asking for a particular IP address (or 0.0.0.0 if you are certain the host will assign the address). If the host does not approve of the choice, it will send a negative acknowledge and in the reply it will indicate which IP address the client should use. IPCP also allows you to negotiate IP compression. However, embedded systems will rarely want to compress IP data. RFC1332 defines the IPCP protocol.

Once the link is set up, IP packets are wrapped with PPP information and then further wrapped with the HLDC envelope. The resulting package is sent via modem to the other side of the connection.

PPP-Related RFCs

RFC Number	Contents
STD2	Assigned protocol numbers
1661	PPP
1662	PPP in HLDC-like framing
1663	Numbered mode PPP option (reliable delivery of PPP packets)
1172	PPP initialization options
1332	IPCP Protocol

INDEX

Note: Boldface numbers indicate illustrations.

About the Author

Al Williams is an electrical engineer best known for writing a wide range of programming books, from *DOS 5: A Developer's Guide* and *Commando Windows Programming*, to *Windows 2000 Systems Programming Black Book* and *MFC Black Book*. A frequent contributor to *Dr. Dobb's Journal*, Java columnist for *Web Techniques*, and formerly the C++ columnist for *Visual Developer* magazine, he is also the author of the recent book *Microcontroller Projects with Basic Stamps* and of structured teaching courses on SX programming, which many universities have selected to satisfy ABET requirements.